CHEMICAL ENGINEERING METHODS AND TECHNOLOGY SERIES

PHOTOCATALYSIS ON TITANIA-COATED ELECTRODE-LESS DISCHARGE LAMPS

Chemical Engineering Methods and Technology Series

Flame Retardants: Functions, Properties and Safety
Paulo B. Merlani
2009. ISBN: 978-1-60741-501-5

Handbook of Membrane Research: Properties, Performance and Applications
Stephan V. Gorley (Editor)
2009. ISBN: 978-1-60741-638-8

Non-Ionic Surfactants
Pierce L. Wendt and Demario S. Hoysted (Editors)
2009. ISBN: 978-1-60741-434-6

Handbook of Hydrogels: Properties, Preparation & Applications
David B. Stein (Editor)
2009. ISBN: 978-1-60741-702-6

Handbook of Hydrogels: Properties, Preparation & Applications
David B. Stein (Editor)
2009. ISBN: 978-1-61668-167-8
(Online Book)

Treatment of Tannery Effluents by Membrane Separation Technology
Sirshendu De, Chandan Das, and Sunando DasGupta
2009. ISBN: 978-1-60741-836-8

Asphaltenes: Characterization, Properties and Applications
Jeremy A. Duncan (Editor)
2010. ISBN: 978-1-60741-453-7

Asphaltenes: Characterization, Properties and Applications
Jeremy A. Duncan (Editor)
2010. ISBN: 978-1-60741-453-7
(Online Book)

Chemicals and Radioactive Materials and Human Development
Rajiv K. Sinha, Rohit Sinha, Shanu Sinha
2010. ISBN: 978-1-61668-145-6

Chemicals and Radioactive Materials and Human Development
Rajiv K. Sinha, Rohit Sinha, Shanu Sinha
2010. ISBN: 978-1-61668-145-6
(Online Book)

Bulk Materials: Research, Technology and Applications
Teodor Frías and Ventura Maestas (Editors)
2010. ISBN: 978-1-60692-963-6

Chromatography: Types, Techniques and Methods
Toma J. Quintin (Editor)
2010. ISBN: 978-1-60876-316-0

Biomedical Chromatography
John T. Elwood (Editors)
2010. ISBN: 978-1-60741-291-5

Biomedical Chromatography
John T. Elwood (Editors)
2010. ISBN: 978-1-60741-291-5
(Online Book)

Bisphenol A and Phthalates: Uses, Health Effects and Environmental Risks
Bradley C. Vaughn (Editors)
2010. ISBN: 978-1-60741-701-9

Bisphenol A and Phthalates: Uses, Health Effects and Environmental Risks
Bradley C. Vaughn (Editors)
2010. ISBN: 978-1-60741-701-9
(Online Book)

Environmentally Harmonious Chemistry for the 21st Century
Masakazu Anpo and K. Mizuno (Editors)
2010. ISBN: 978-1-60876-428-0

Green Composites: Properties, Design and Life Cycle Assessment
François Willems and Pieter Moens (Editors)
2010. ISBN: 978-1-60741-301-1

Photocatalysis on Titania-coated Electrode-less Discharge Lamps
Vladimír Církva and Hana Žabová
2010. ISBN: 978-1-60876-842-4

Perovskites: Structure, Properties and Uses
Maxim Borowski (Editors)
2010. ISBN: 978-1-61668-525-6

Perovskites: Structure, Properties and Uses
Maxim Borowski (Editors)
2010. ISBN: 978-1-61668-525-6
(Online Book)

Fluid Phase Behavior of Systems Involving High Molecular Weight Compounds and Supercritical Fluids
Pedro F. Arce and Martín Aznar
2010. ISBN: 978-1-61668-310-8

Preparation of Thin Film Pd Membranes for H2 Separation From Synthesis Gas and Detailed Design of a Permeability Testing Unit
M. Bientinesi, L. Petarca
2010. ISBN: 978-1-60876-538-6

CHEMICAL ENGINEERING METHODS AND TECHNOLOGY SERIES

PHOTOCATALYSIS ON TITANIA-COATED ELECTRODE-LESS DISCHARGE LAMPS

VLADIMÍR CÍRKVA
AND
HANA ŽABOVÁ

Nova Science Publishers, Inc.
New York

For permission to use material from this book please contact us:
Telephone 631-231-7269; Fax 631-231-8175
Web Site: http://www.novapublishers.com

LIBRARY OF CONGRESS CATALOGING-IN-PUBLICATION DATA

Církva, Vladimír, 1968-
Photocatalysis on titania-coated electrode-less discharge lamps / Vladimír Církva and Hana Žabová.
p. cm.
Includes bibliographical references and index.
ISBN 978-1-60876-842-4 (softcover)
1. Electric discharge lighting. 2. Photocatalysis. 3. Titanium dioxide--Electric properties. 4. Titanium dioxide films. I. Žabová, Hana, 1982- II. Title.
TK4371.C57 2009
621.32'5--dc22

2009049855

Published by Nova Science Publishers, Inc. † New York

CONTENTS

PREFACE

This book demonstrates outstanding improvement of photocatalytic degradative efficiency by the coupling of microwave (MW) radiation with the titania-coated electrodeless discharge lamps (EDLs). Titania thin film is used as the photocatalyst due to its superior characteristics. The EDL as a novel light source generates efficiently UV/Vis radiation when placed into a MW field. Thin nanoporous titania films are prepared by dip-coating of EDL via a sol-gel method using titanium isopropoxide, or titanium *n*-butoxide, acetylacetone, and a transition metal acetylacetonate. The films are characterized through XRD, Raman spectroscopy, XPS, SEM, AFM, and UV/Vis. The integration of dopants into the sol during gelation process is performed and some transition metal doped titania films with M^{n+} (M = Fe, Co, Ni, V, Cr, Mn, Zr, Ag) are prepared. Photocatalytic activity of titania films is followed by decomposition of Rhodamine B in water. The photocatalytic reactions on the EDL thin films are carried out at relatively high concentration (0.1 mol l^{-1}) of mono-chloroacetic acid (MCAA) in the MW field, and the MCAA is efficiently decomposed to HCl, CO_2, and H_2O. The effect of operational parameters is reported for both batch photoreactor (number of coating cycles for EDL, light intensity, initial pH value, and H_2O_2 dosage) and for continuous-flow set-up (flow rate, number of titania-coated EDLs, reaction temperature, and air bubbling). The MCAA decomposition is enhanced in an alkaline solution and in the presence of H_2O_2, and significantly enhanced by increasing the intensity of UV light. Studies revealed that reaction temperature and light intensity of the EDLs depend inversely on the flow rate, but that the 366 nm line intensity of EDL is directly proportional to the reaction temperature. The photodegradation of MCAA is enhanced by heating and significantly enhanced by air bubbling of the reaction mixture in the glass reservoir at laboratory temperature. The photocatalytic efficiency increases with the number of titania-coated EDLs. Compared to the pure TiO_2 film, the UV/Vis spectra of V, Zr, and Ag-doped TiO_2 show significant absorption in the visible region, and hence the photocatalytic degradation of MCAA is increased. The best apparent degradation rate constant (0.0125 min^{-1}), which is higher than that on the pure TiO_2 film by 1.7 times, is obtained with the Ag(3%)/TiO_2 photocatalyst. The effect of doping level of vanadium acetylacetonate on the photocatalytic efficiency of the V-doped TiO_2 is also determined.

Keywords: *Microwave assisted photocatalysis, Electrodeless discharge lamp, Mono-chloroacetic acid, Rhodamine B, Nanoporous titania thin film, Transition metal ion doped titania, Sol-gel, Photoactivity, Batch and continuous-flow photocatalytic reactor*

Chapter 1

INTRODUCTION

Photocatalysis is an efficient, attractive and clean technology for pollution abatement in water under mild conditions [1]. Titanium(IV) oxide, known as titania, is the archetypical photocatalytic material since it is endowed with an inherent photocatalytic activity. Moreover, it is inexpensive, very chemical and thermal stable, nontoxic and available in large amounts [2]. For better recovery and reuse, titania can be prepared in immobilized form as a thin film on glass substrate as a catalyst support. A major advantage here is that the reaction products and photocatalyst do not have to be separated unlike in the cases with powder or colloidal suspensions (slurry) of the photocatalyst [3]. However, the immobilized films produced by common sol-gel methods exhibit relatively low photocatalytic activity because of their low surface area and small film thickness. Likewise, the fast recombination rate (mean e^-/h^+ lifetime is about 30 ns) of photoinduced electron-hole pairs hinders the utilization of photocatalysis.

The application of microwave (MW) energy on photocatalysis enables to solve the above mentioned problems. MW energy is a non-classical energy source, with photoactivation, ultrasound, high pressure, electrochemistry, or plasma discharge. MW activation increases the efficiency of many processes and can simultaneously reduce formation of the byproducts obtained from conventionally heated reactions. Since first reports of the use of MW heating to accelerate organic chemical transformations [4], numerous articles have been published on the subject of MW-assisted synthesis and related topics, microwave chemistry, has certainly became an important field of modern organic chemistry [5]. Chemical processes performed under the action of MW radiation are believed to be affected in part by superheating, hot-spot formation, and polarization [6].

The existence of a specific non-thermal MW effect has been a matter of controversy during the past years [5,7].

MW heating has already been used in combination with some other non-conventional activation processes. Such a combination might have a synergic effect on reaction efficiencies or, at least, enhance them by summing up the individual effects. Application of MW radiation to ultrasound-assisted chemical processes has been recently described [8-10]. There have also been attempts to affect electrochemical [11], photochemical [12,13] and photocatalytic reactions [14].

The energy of MW radiation (E = 0.98 J mol^{-1} at ν = 2.45 GHz) is considerably lower than that of UV/Vis radiation (E = 600-170 kJ mol^{-1} at λ = 200-700 nm), thus insufficient to disrupt the bonds of common organic molecules. We therefore assume that, essentially, UV/Vis is responsible for the electron-hole generation in a photocatalyst, and MW radiation subsequently affects the course of the subsequent electron-hole reactions. Moreover, the generation of hydroxyl radicals can be also increased by MW irradiation [15].

The objective of MW-assisted photocatalysis is frequently, but not necessarily, connected to the titania-coated *electrodeless discharge lamp* (EDL) as a novel light source which generates efficiently UV/Vis radiation when placed into a MW field. Review demonstrates outstanding improvement of degradative efficiency by the coupling of MW radiation with EDLs to the photocatalytic degradation on the thin titania films.

Chapter 2

UV/VIS DISCHARGES IN ELECTRODELESS LAMPS

The *electrodeless discharge lamp* (EDL) [16] consists of a glass envelope tube filled with an inert gas and an excitable substance and it is sealed under a lower pressure (2-20 Torr) of a noble gas. MW field can trigger gas discharge causing the emission of UV/Vis radiation. This phenomenon has been studied for many years and was well understood in the 1960s [17]. The term "electrodeless" means that the lamps lack the electrodes within the envelope. Meggers [17] developed the first EDL using the mercury isotope ^{198}Hg in 1942 and its earliest application was in absorption spectroscopy [18]. EDL is usually characterized by a higher emission intensity than that of hollow cathode lamps, lower contamination, because of the absence of the electrodes, and a longer lifetime [19]. The lamps were used in various applications as light sources as well as in atomic spectrometers [20].

2.1. THEORY OF THE DISCHARGE

The theory of Hg-EDL operation, as it is currently understood, is shown in Figure 1. Free electrons in the fill (i.e. electrons that have become separated from the environment because of the ambient energy) accelerate as a result of the MW field energy [13]. They collide with the argon atoms and ionize them to release more electrons. The repetitive effect causes the number of electrons to increase significantly over a short period of time, an effect known as an "avalanche".

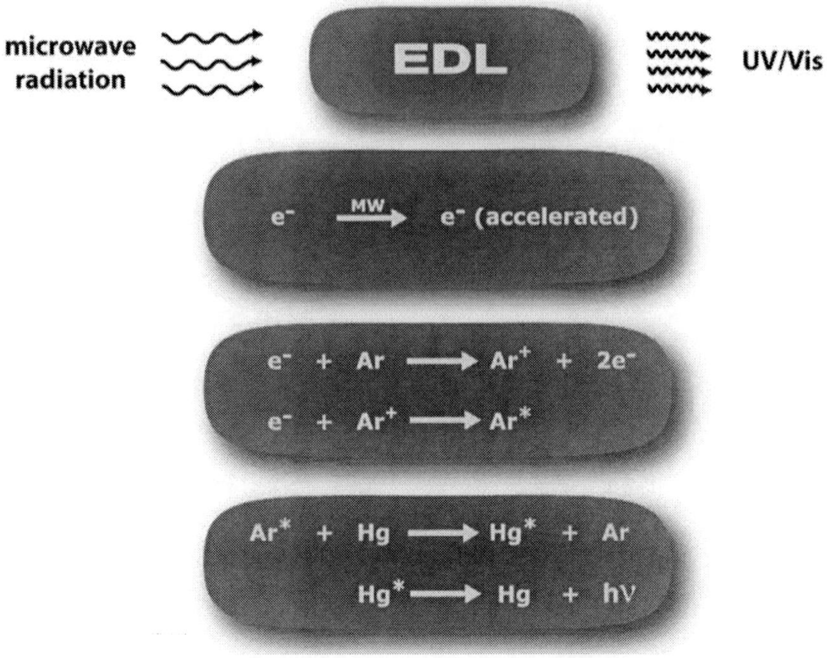

Figure 1. Principle of Hg-EDL operation and the release of emission energy as UV/Vis radiation [13].

The energetic electrons collide with the heavy-atom particles (argon or mercury) present in the plasma, exciting them from a ground state to higher energy levels. The excitation energy is then released as UV/Vis radiation with spectral characteristics which depend on the composition of the envelope. The excited molecular or atomic species in the plasma can emit photons over very broad portion of the electromagnetic spectrum, ranging from x-rays to the IR [21].

2.2. Construction of Electrodeless Lamps

The EDL system is modular and consists of two basic parts, a gas filled bulb and a MW power supply. A typical EDL is made of a quartz or Pyrex tube envelope, which contains an argon and an excitable substance. The envelope material must be impermeable to gases, an electrical insulator, and chemically resistant to the filling compounds at the temperature of operation.

The construction of microwave-excited EDLs is relatively straightforward but there are a number of operating parameters in their preparation which have to be considered in order to produce an intense light source. The desired characteristics and requirements for EDL are high intensity, great stability, long lifetime, and to a lesser extent, low cost and high versatility. In practice, it is very difficult to simultaneously meet all these desired characteristics.

General procedures of EDL construction are available in the literature [22-24]. However, many minor details which are critical for the lamp proper function are often omitted. The investigator who wants to make an EDL is thus faced with a very large amount of information dispersed in the literature, and finds that it is very difficult to reproduce these procedures to develop EDLs having desired properties. An experimental vacuum system for EDL (Hg, HgI_2, Cd, I_2, KI, P, Se, and S) construction has been recently designed by Církva *et al.* (Figure 2) [25]. The Pyrex EDL blank was cleaned in a water-soap mixture, and then washed with distilled water, aqueous 10% hydrofluoric acid, and ethanol. 2.5 μl Hg and a stainless steel thin wire (3 cm) were placed to the EDL blank. The system was flushed with argon and sealed under 20 Torr vacuum. This technique is very simple and enables to prepare EDL in a common chemical laboratory.

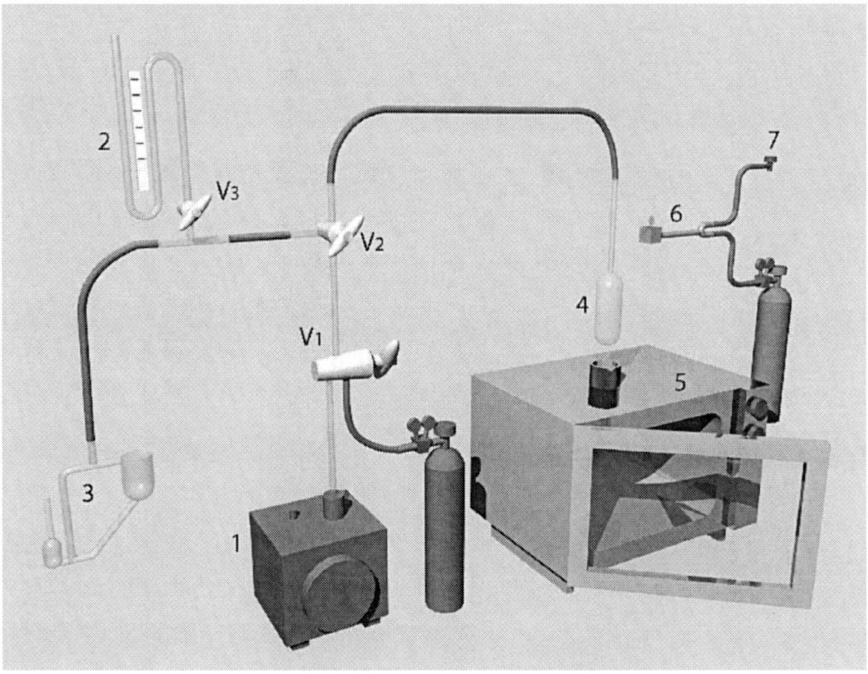

Figure 2. Vacuum system for the EDLs construction: (1) rotary vacuum pump; (2) mercury manometer; (3) tilting-type McLeod pressure gauge; (4) EDL blank; (5) modified microwave oven; (6) glass-working burner; (7) natural gas; V_1-V_3 are stopcocks [25].

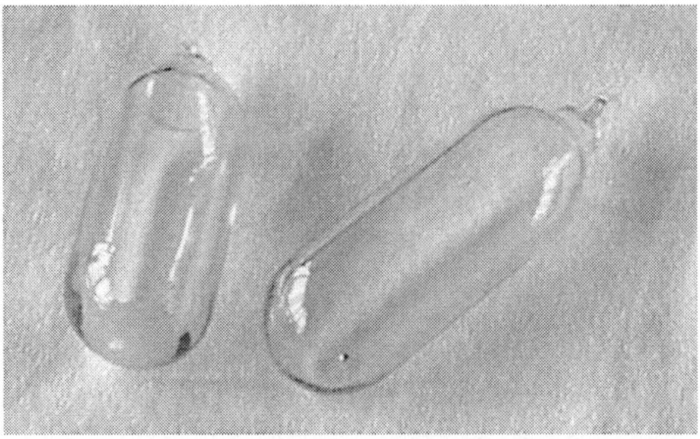

Figure 3. Hg-EDL and S-EDL for photochemical applications.

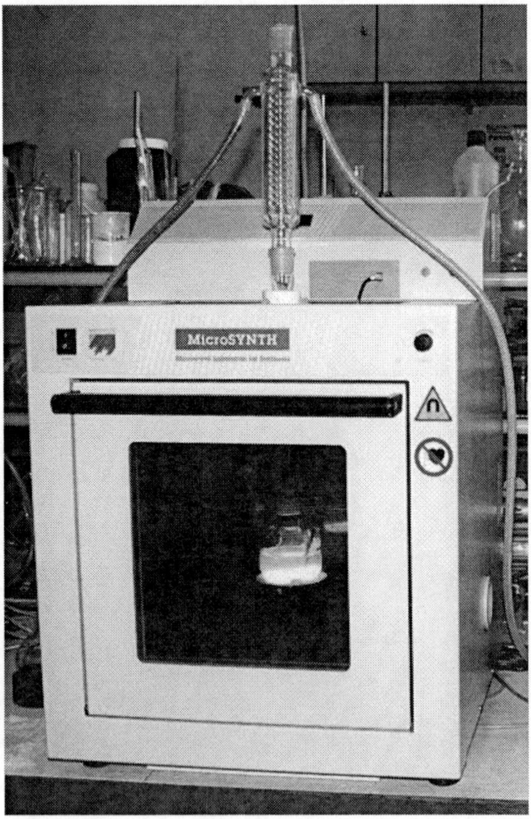

Figure 4. Testing the EDL performance in a Milestone´s MicroSYNTH Labstation [25].

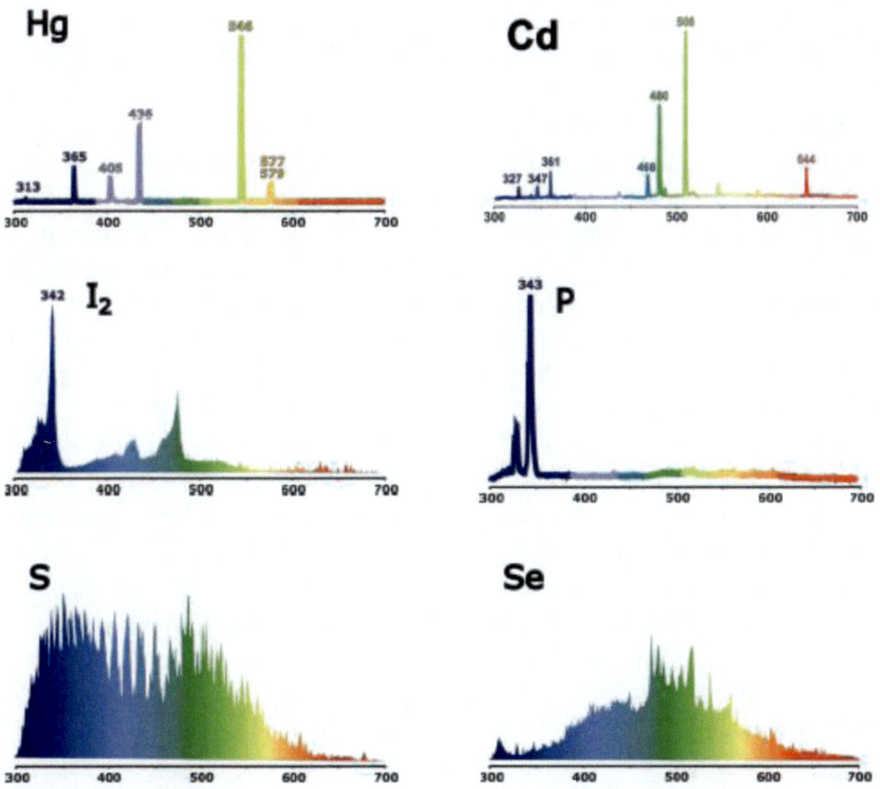

Figure 5. Emission spectra of Hg-, Cd-, I-, P-, S-, and Se-EDL (Pyrex envelope; argon atmosphere) [27].

The EDL length was 50 mm (diameter 20 mm) and the pictures of Hg-EDL and S-EDL are shown in Figure 3. Testing the EDL performance was carried out in order to prepare the lamps for spectral measurements [25]. A typical experimental system for such a testing consisted of a round-bottom flask containing *n*-heptane, equipped with a fiber-optic temperature and spectral probe, a Dimroth condenser, and placed to a MW oven (Figure 4).

2.3. Spectral Characteristics of Electrodeless Lamps

The knowledges of spectral characteristics of EDL are clearly essential for planning the microwave-assisted photocatalytic experiments. The right choice of

EDL envelope and fill material can be very useful in planning an efficient course of the photocatalytic process without necessity of filtering off the undesirable part of the UV radiation by other tools, such as glass or solution filters or monochromators [26,27]. While atomic fills usually furnish *line emission spectra*, molecular fills give *continuous emission bands* [27]. The total emission output of Hg-EDL in the region of 200-700 nm is approximately the same as that of the electrode lamp with the same power input [28]. The distribution of radiation is, however, markedly different, as a result of a much higher Hg pressure and the greater number of atoms that are present in the plasma. EDL emits over three times as much UV and over a half as much IR as a conventional lamp [29]. Addition of some material had a very significant effect on the spectral distributions of EDL [28].

Müller, Klán, and Církva have reported [27] the emission characteristics of various EDL containing different fill material (such as, Hg, HgI_2, Cd, I_2, KI, P, Se, or S) in the region of 250-650 nm. While distinct *line emission peaks* were found for the mercury, cadmium, and phosphorus fills (Figure 5), the iodine-, selenium-, and sulfur-containing EDL (Figure 5) emitted *continuous bands*. Sulfur-containing EDL has been proposed to assist phototransformations that are of environmental interest because the emission flux is comparable to solar terrestrial radiation. In addition, the EDL spectra could be easily modified by the choosing a suitable EDL envelope glass material, temperature, MW output power, or solvent according to the needs of a photochemical experiment [26]. Spectral measurements of prepared EDLs (**light intensity in $\mu W.cm^{-2}$**) were carried out on the spectrometer AVS-S2000 with AvaSoft software package and the spectrometer USB2000 with an optical fiber probe and operating software package OOIrrad-C (Ocean Optics) [25-27].

2.4. Performance of Electrodeless Lamps

The performance of EDLs depends strongly on many preparation and operating parameters:

a) *The inert gas*. The arc chamber contains a buffer noble gas (usually Kr, Xe, or Ar) which is inert to the extent that it does not adversely affect the lamp operation. Helium has a higher thermal conductivity than other noble gases and, therefore, higher thermal conduction loss is observed [30]. The inert gas easily ionizes at low pressure but its transition to the thermal arc is slower and the lamp requires a longer warm-up time.

Ionization is more difficult at higher pressures and it requires a higher input power to establish the discharge. In general, the pressure of filling gas was recommended to hold between 0.266-2.66 kPa (2-20 Torr), at the operating temperature is usually much higher (5-20 atm) than that of a conventional electrode lamp. Utilizing argon was considered to be the best compromise between high EDL radiance and long lifetime. Air cannot be used, due to the quenching properties in microwave plasma just like water vapors.

b) *The choice of the fill material* initiating the discharge is very important. Together with a standard mercury fill it is often desirable to incorporate an additive in the fill material that has a low ionization potential and a sufficient vapor pressure (Cd, S, Se, Zn) [31,32]. One category of low-ionization-potential materials is the group of alkali metals or their halides (LiI, NaI) but some other elements, such as Al, Ga, In, Tl [33,34], Be, Mg, Ca, Sr, La, Pr, or Nd [30,35,36], can be used. Other metal-containing compounds have been utilized to prepare EDL, including amalgams of Cd, Cu, Ag, and Zn. Multi-element EDL have been prepared using combinations of elements (e.g., Li-Na-K, As-Sb, Co-Ni, Cr-Mn, Bi-Hg-Se-Te, Cd-Zn, Ga-In, Se-Te) [37]. The spectral output from each individual element is very sensitive to temperature [38]. It has been found that no inter-element interferences occur in the lamp.

c) *Temperature of the lamp.* Operation at a high power or high temperatures can increase the emission intensity but, at the same time, reduce the lamp lifetime and lead to a broadening of the atomic line profile due to self-absorption and self-reversal effects. It was found that the optimum operating temperature for the mercury filling is 42 °C (for 254 nm line) [39]. The output is reduced when the temperature is beyond optimum [26].

d) *The dimensions of the lamp envelope* are based on the discovery that the volume of Hg is critical for the effective UV operation [40]. Higher Hg pressures result in the need to use higher MW power levels. To focus the MW field efficiently into EDL, a special Cd low-pressure lamp with a metal antenna (a molybdenum foil) was developed for experiments in MW-absorbing liquids [41]. High quality quartz is the most widely used lamp envelope material but early EDL manufacturing used glass, Vycor, or Pyrex [22].

e) *The frequency and intensity* of electromagnetic energy is determined by the type of a device. MW energy is widely used for the excitation of EDL because it is generally more efficient than radiofrequency energy for the generation of intense light. MW radiation for the excitation of gas discharges is usually generated by a fixed frequency (2.45 GHz) magnetron oscillator.

Chapter 3

THIN TITANIA FILMS ON ELECTRODELESS LAMPS

Titanium(IV) oxide is a polymorphic substance that is present in three crystalline varieties: rutile, anatase, and brookite, with different properties and structure. Although anatase seems to be the stable form of titanium dioxide at room temperature, the polymorphic brookite transforms to rutile at high temperature (up to 700 °C) [42,43].

Photon absorption by TiO_2 results in the promotion of an electron from the valence band to the conduction band of the titania, leaving behind a "hole" (i.e. electron vacancy) in the valence band, provided the photon has an energy at least equal to that of the bandgap of the photocatalyst. The bandgaps for anatase and rutile being 3.2 eV and 3.0 eV, corresponding to wavelengths 385 nm and 410 nm, respectively, UV light is needed for photoexcitation. The electron hole pairs can either recombine or participate in chemical reactions with surface/adsorbed species. Oxidation of water/hydroxide ion by the valence-band hole (h^+_{VB}) can produce the hydroxyl radical OH•. The conduction-band electron (e^-_{CB}) can react with molecular oxygen to form the superoxide radical-anion, O_2^{-}•, which can be involved in further reactions. In addition, h^+_{VB} and e^-_{CB} can react directly with adsorbed pollutants. The radicals thus formed, dioxygen and water can participate in further reactions, resulting ultimately in mineralization of the organic pollutants.

3.1. Preparation of the Thin Titania Films

Nanostructured titania has been prepared by the chemical methods such as *the hydrothermal technique, chemical vapour deposition (CVD)* [44,45] or *sol-gel route*; physical methods for example *magnetron sputtering* and plasma-chemical methods [46,47] such as *plasma chemical vapour deposition (PCVD), plasma assisted chemical vapour deposition (PACVD)* or *plasma enhanced chemical vapour deposition (PECVD)*. Titania, in anatase and rutile form, is obtained through the precipitation method which implies a careful hydrolysis of the titanium sulfate or titanium chloride and subsequent dehydration of the formed titania. The preparations conditions are particularly decisive in the transformation temperature of anatase to rutile and it has been found that the precipitation method produces anatase even though it is used a sulfate or chloride. There is therefore a problem with the hydrolysis, which originates from the formation of ionic species that are produced in the precipitation and which are related to a great extent with the pH [48,49].

Hydrothermal synthesis, in which chemical reactions can occur in aqueous or organic media under the self-produced pressure at low temperature (usually lower than 250 °C), can be a suitable means of titania films preparation. This automatically raises the effective boiling point of the solvent, which can also solve the problems due to the limitations of their rather low boiling points. This technique is also called solvothermal, while in the special case of the solvent being water, often called hydrothermal. The solvothermal treatment could be useful to control grain size, particle morphology, crystalline phase and surface chemistry by regulating solution composition, reaction temperature, pressure, property of solvent, additives and aging time [50,51].

Sol-gel route is one of the most successful techniques for preparing nanosized metallic oxide materials with high photocatalytic activities. By tailoring the chemical structure of primary precursor and carefully controlling the processing variables, nanocrystalline products with very high level of chemical purity can be achieved [52]. In sol-gel processes, TiO_2 is usually prepared by the reactions - hydrolysis and polycondensation of titanium alkoxides, $Ti(OR)_n$ to form oxopolymers, which are then transformed into an oxide network [53-55]. The overall reaction is usually written as follows [56]:

$$Ti(OR)_n + 0.5n\ H_2O \rightarrow TiO_{n/2} + 4\ ROH$$

The condensation is usually accomplished by gelization and calcination. The condensation pulls together the constitute particles of the gel into a compact mass, thus building up the metal oxide crystal. The calcination temperature is especially important for removing the organic molecules from the final products. The sol-gel derived precipitates are amorphous in nature, requiring further heat treatment at a high temperature to induce crystallization. The calcination process gives rise to particle agglomeration and grain growth and hence may induce undesirable phase transformation [57-59]. Therefore, development of a process without the calcination step for crystallization may be more favorable. For example the extractions with subcritical water and/or supercritical carbon dioxide have been employed as the steps limiting the content of organics [60]. It is also possible to prepare numerous new materials with a regular and controlled morphology on the nano- and microscale. The sol-gel method within templates of surfactants assemblies organized as reverse micelles is effective strategy for generation of uniform metal oxide nanoparticles finalized as thin films [61]. In this process the alkoxide hydrolyses inside a reverse micelle with a limited amount of water. Then the polycondensation step could proceed simultaneously and is highly competitive. Compared to other methods the surfactant mediated sol-gel provides a good control of the hydrolysis rate [62].

The titania photocatalyst on EDL can be prepared using a sol-gel method:

(a) from titanium(IV) isopropoxide: The titania sol was prepared by hydrolysis of titanium(IV) isopropoxide according to the method described by Kluson et al. [60]. The hydrolysis was carried out in the reverse micelles of Triton X-100 in cyclohexane. The molar ratios of initial compounds were 1:1:1 (water : Triton X-100 : titanium isopropoxide) and volume ratio Triton X-100 : cyclohexane was 0.45.
(b) from titanium(IV) butoxide: Titanium(IV) butoxide was dissolved in acetylacetone and used as a molecular precursor of TiO_2. The resulting reaction originated a solution into which the solvent ethyl alcohol was added gradually. The volume ratio of these reactants was 1:1:1. Then the water used for hydrolysis was added dropwise under mechanical stirring and molar ratio titanium butoxide/water was 0.1.

As in the case of all surface-finishing techniques, it is also important to maintain a high level of surface cleanliness to ensure good adhesion between the substrate and the surface coating. Prior to the film deposition, the support (EDL) was thoroughly cleaned in a water-soap mixture, rinsed with distilled water, soaked in a solution of HCl (1.0 mol l^{-1}), rinsed with water and ethanol, respectively. Then the coatings of pretreated supports were made in the dip-coating machine with withdrawal rate 6 cm min^{-1}.

Figure 6. Titania-coated Hg-EDL.

The dip-coating process is divided in five stages, immersion, start-up, deposition, evaporation and drainage. In this method the support is slowly dipped into and withdrawn from a tank containing the gel, with a uniform velocity, in order to obtain a uniform coating. The films were then dried at room temperature for one hour and finalized by thermal treatment at 500 °C for 2 h (2 °C/min).

The titania-coated Hg-EDL was at last completed (Figure 6) and prepared for following *microwave photocatalytic* experiments.

3.2. Transition Metal Doped Thin Titania Films

Despite the positive attributes of titania, the main drawback associated with its use is that most of the activated charge carriers will undergo recombination before reaching the surface to interact with adsorbed molecules [63]. Another challenge of utilizing titania for environmental remediation is that pristine TiO_2 is photoactive only under UV irradiation with an adsorption edge wavelength of typically less than 388 nm (i.e. 3.2 eV band gap), which accounts for less than 5% of solar light energy. This thus creates difficulty of its potential applications [64]. The recombination of photo-induced electrons and holes can occur very quickly (mean lifetime of an e^-/h^+ pair is about 30 ns). If the recombination occurs too fast, then there is not enough time for any other chemical reactions to occur [65]. In fact, 90% of the generated carriers are lost within ns of their generation, leading to low photoactivity of titania [66]. To circumvent these two particular limitations, a number of strategies have been proposed to improve the light absorption features and lengthen the carrier lifetime characteristics of TiO_2. In recent years, impurity doping, dye sensitization [67] or semiconductors coupling [68] have been widely performed in order to improve photoactivity. Because transition metal elements have many valences and because trace transition metal ions doped in the TiO_2 matrix can be superficial potential traps of photo-generated e^-/h^+ pairs. Then the processes lengthen the lifetime of electrons and holes and increase the photocatalytic activity [69].

The dopants are usually confined as very dispersed species to the surface and/or to a few top layers of TiO_2 particles, due to the moderate calcination temperatures used in the conventional preparations. Doping TiO_2 with transition metal ions [70-74] has been frequently attempted not only to retard the fast charge pair recombination but also to enable visible light absorption by providing defect states in the band gap. In the former case, the UV-excited electrons in the conduction band (CB) or holes in the valence band (VB) are trapped in the defect sites by retarding the recombination and subsequently enhancing the interfacial charge transfers. On the contrary, the intraband states may also serve as recombination centers to decrease the overall photocatalytic activity [75]. In the latter case, the electronic transitions from VB to defect states or from defect states to CB can be allowed under sub-band-gap illumination to show visible activity [76].

A comparison among the results reported in the literature for doped samples of TiO_2 obtained from various preparations is not easy because the experimental conditions under which the reactions are carried out and the preparation methods of the samples are usually different. It is well known that bare titania show very different physicochemical and electrochemical properties. It has been shown that the photocatalytic activity of TiO_2 can be influenced by its crystal structure, surface area, size distribution, porosity, band gap and surface hydroxyl group density [77]. Nevertheless, it is worth pointing out that the photoactivity cannot be straightforwardly related to only few properties because it depends on all of them. The success of doping operation depends on the site where the dopant ion is placed. Dopant ions will impact the properties in a different way if they are placed substitutionally on regular bulk sites, on interstitial positions, or segregated to boundaries [78].

The expensive ion implantation method [79] has been used to obtain photocatalysts more active than the corresponding bare titania. In this case ions are believed to be present in substitutional positions in the lattice of TiO_2. Since the sol-gel method is a solution process, it has all the advantages of wet chemical process such as control of stoichiometry, doping of desired amount of transition metal ions and fine dispersion of the dopant and titanium source. Many attentions have been paid to doping the titanium(IV) oxide with transition metals such as iron [80,81], cobalt [82,83], nickel [84], manganese [85], chromium [86,87], vanadium [88,89], copper [90], molybdenum [91], tungsten [92], zinc [93], zirconium [94,95] or silver [96-98]. The photodegradation of various substrates is then a complex function of the doping ion type, concentration and the microstructural characteristics of the catalysts. There usually exists a critical doping concentration at which further increase will result in the charge carrier recombination, thus lowering the photoactivity of prepared samples [99]. It has been also found that doping of the TiO_2 crystals with transition metal ions in some cases however reduced the photooxidation rates compared to the pristine TiO_2 [100].

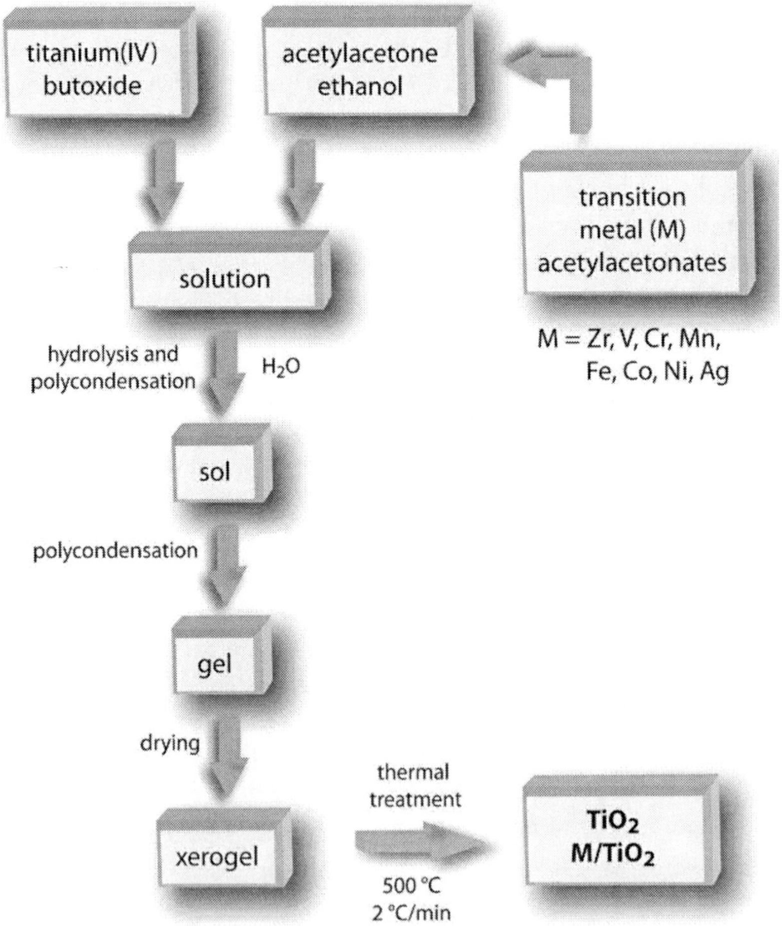

Figure 7. Flow chart for the sol-gel processing of titania-doped thin films.

Transition metals titania-doped (M/TiO_2) thin films have been prepared by sol-gel method using titanium(IV) butoxide and transition metal acetylacetonate as precursors. For the preparation of M/TiO_2 sol, transition metal acetylacetonate (0.1-0.95 g) was dissolved in a 10 ml of acetylacetone. After that, titanium(IV) butoxide was added followed by the addition of 0.1 ml of concentrated nitric acid and 10 ml of ethanol. Finally, 4 ml of water was added dropwise. In order to obtain a homogeneous mixture of M/TiO_2, the solution was stirred vigorously for 2 h. Figure 7 shows the flow chart for the preparation of transition metals titania-doped thin films. Typical sol was prepared for various transition metals (M = Zr, V, Cr, Mn, Fe, Co, Ni, and Ag) and various wt.% of transition metal concentrations (1, 3, 5, and 9 wt.%). The films of M/TiO_2 were deposited on substrates (glass slide or EDL) by dip-coating technique and then dried at room temperature for one hour and calcined at 500 °C for 2 hours (heating rate 2 °C/min).

3.3. Characterization of Thin Titania Films

The prepared titania and titania-doped (M/TiO_2) thin films were characterized by several techniques: X-ray diffraction (XRD), Raman spectroscopy, X-ray photoelectron spectroscopy (XPS), scanning electron microscopy (SEM), atomic force microscopy (AFM), and UV/Vis absorption spectroscopy.

3.3.1. X-ray Diffraction (XRD)

The titania and titania-doped thin films were characterized for crystal phase identification by XRD using a XRD-7 (Rich. Seifert and Co., Freiberg, Germany) with CuKα radiation ($\lambda = 0.15418$ nm) at $2\Theta = 20\text{-}80°$. The average diameters of the TiO_2 particles (d_{TiO_2}) were obtained by means of the Scherrer equation [101]:

$$d_{TiO_2} = \frac{k \cdot \alpha}{\beta \cdot \cos\Theta} \qquad (1)$$

where $\boldsymbol{\beta}$ is line broadening ($\beta = \beta_s - \beta_0$, where β_s and β_0 are the half widths of the XRD peaks of the sample and of the standard), $\boldsymbol{k}$ related to the crystallite shape ($k = 0.9$), and $\boldsymbol{\alpha}$ and $\boldsymbol{\Theta}$ are the radiation wavelength and Bragg angle, respectively.

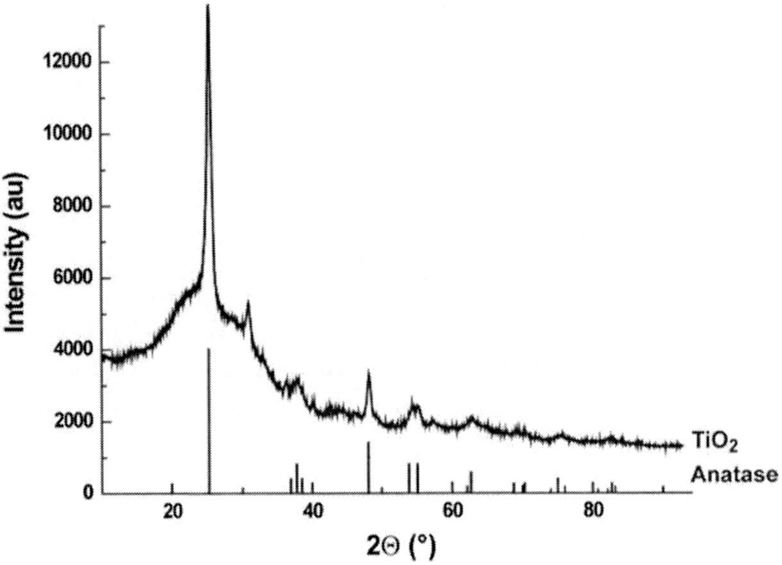
12000
10000
8000
6000
4000
2000
0
Intensity (au)
20
40
60
80
2Θ (°)
TiO2
Anatase

a.

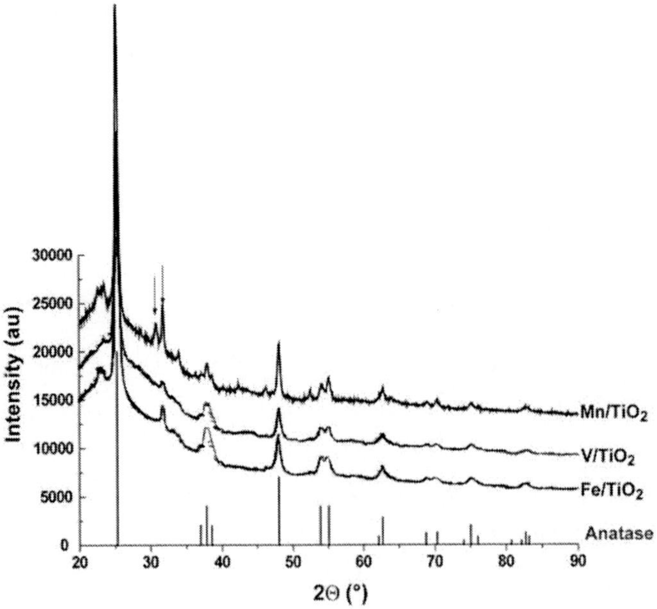
30000
25000
20000
15000
10000
5000
0
Intensity (au)
20
30
40
50
60
70
80
90
2Θ (°)
Mn/TiO2
V/TiO2
Fe/TiO2
Anatase

b.

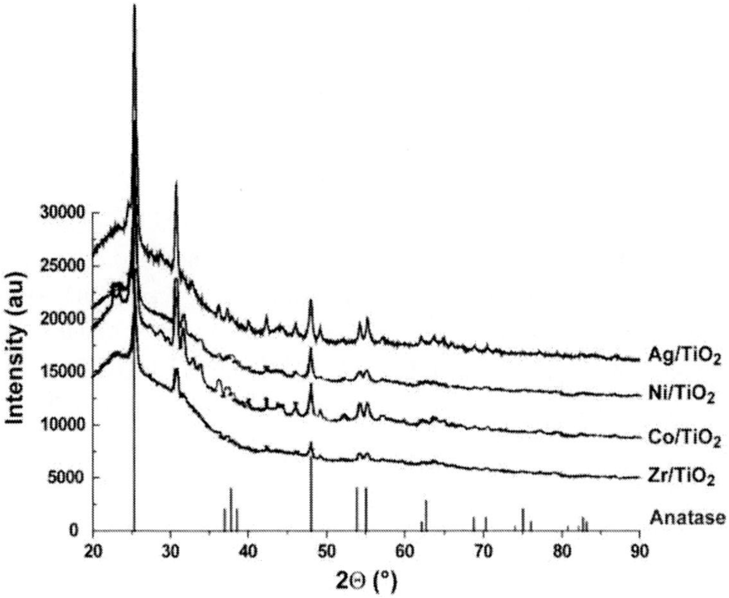

c.

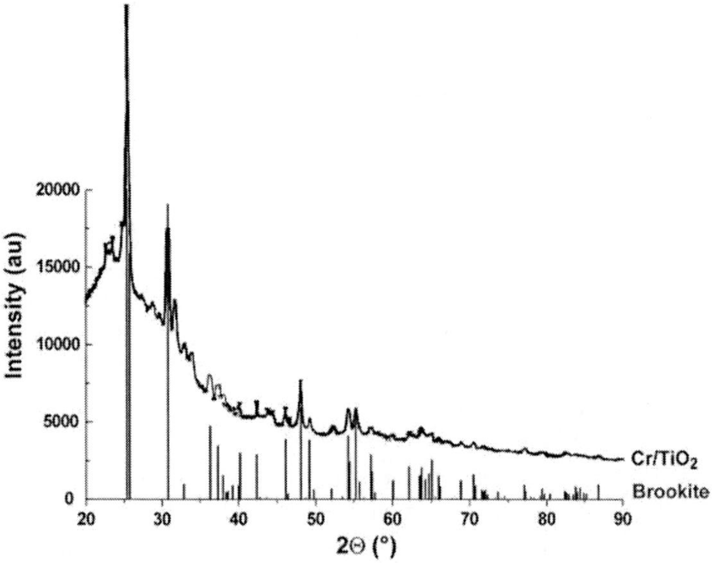

d.

Figure 8. X-ray diffraction patterns of pure TiO_2 (a) and M/TiO_2 (b-d) (M = Mn, V, Fe, Ag, Ni, Co, Zr, and Cr).

The XRD spectrum of the thermally treated pure TiO_2 thin film is shown in Figure 8a. Diffractions that are attributable to anatase crystal phase are clearly detectable in all materials. Figure 8b depicts the XRD patterns of Fe/TiO_2, V/TiO_2 and Mn/TiO_2 samples. The diffractograms of Zr/TiO_2, Co/TiO_2, Ni/TiO_2 and Ag/TiO_2 were displayed in Figure 8c. In all samples, the main diffraction line corresponds to anatase and small lines for brookite phase were also depicted. Other reflections were assigned to triclinic phase of Ti_xO_y. The main diffraction line corresponds to Ti_4O_7 at $2\Theta = 31.74°$ (arrows in Figure 8-b). The XRD pattern of Cr/TiO_2 sample (Figure 8d) shows a mixture of two crystalline structures, a large predominance of the tetragonal anatase phase and the substantial amount of orthorhombic brookite phase were present. All samples feature high microstress that contribute to the broadening of reflections and the average particle sizes of all samples were determined about 20 nm.

3.3.2. Raman Spectroscopy

Raman spectra were recorded from 25-1000 cm^{-1} using a Dispersive Raman Spectrograph (Nicolet Almega XR; Thermo Fisher Scientific, Waltham, USA) and an argon ion laser operating at 473 nm. The spectral bandpass was 100 μm and an uncoated 50x objective was used for all microsampling.

Anatase is tetragonal and belongs to the space group D_{4h}^{19} and Z = 4 [TiO_2]. According to the factor group analysis, the 10 optical modes have the representation [102]: $1A_{1g}+2B_{1g}+3E_g+1A_{2u}+1B_{2u}+2E_u$. The modes A_{1g}, B_{1g} and E_g are Raman active and the modes A_{2u} and E_u are infrared active. The B_{2u} mode is inactive both in the Raman and infrared spectra. Raman spectra of synthetic anatase powder [102] exhibit characteristic bands assigned as A_{1g} (513 cm^{-1}), B_{1g} (399, 519 cm^{-1}) and E_g (144, 197, 639 cm^{-1}). Raman spectrum of undoped TiO_2 has been shown in Figure 9-a and well-resolved Raman peaks were observed at 141 cm^{-1} (E_g), 197 cm^{-1} (E_g), 398 cm^{-1} (B_{1g}), 515 cm^{-1} (B_{1g}) and 640 cm^{-1} (E_g) in the spectra, indicating that anatase is the predominant phase. No effects of dopant and doping concentration were observed on the position of peaks in Raman spectra (Figure 9-b), it means that anatase structure is retained after doping and dopants may occupy the substitutional sites in the host lattice. The two different phases of the films (anatase and brookite) were further confirmed by Raman spectra, as revealed by XRD. Brookite is orthorhombic with space group D_{2h}^{15} and Z=8 [TiO_2]. Factor group analysis indicates a total of 69 optical modes with the following irreducible representation of normal modes [103]: $9A_{1g}+9B_{1g}+9B_{2g}+9B_{3g}+9A_{1u}+8B_{1u}+8B_{2u}+8B_{3u}$.

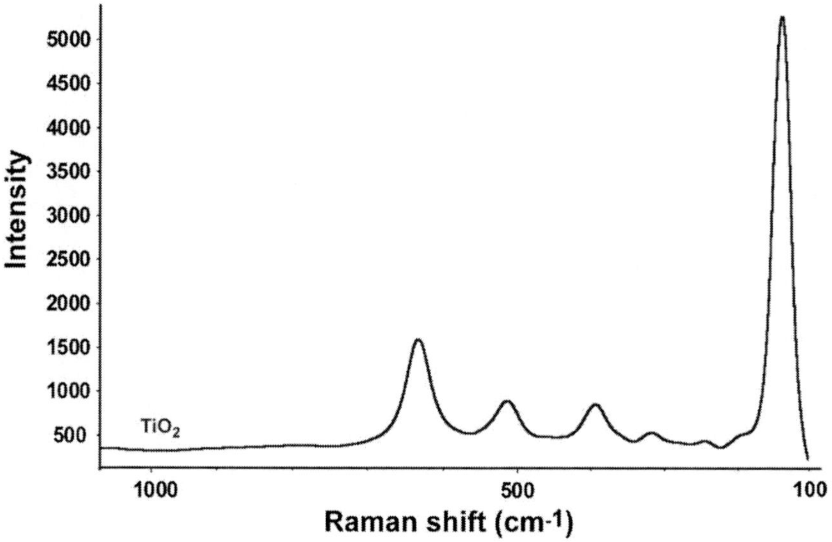
5000
4500
4000
3500
3000
2500
2000
1500
1000
500
Intensity
TiO2
1000
500
100
Raman shift (cm-1)

a.

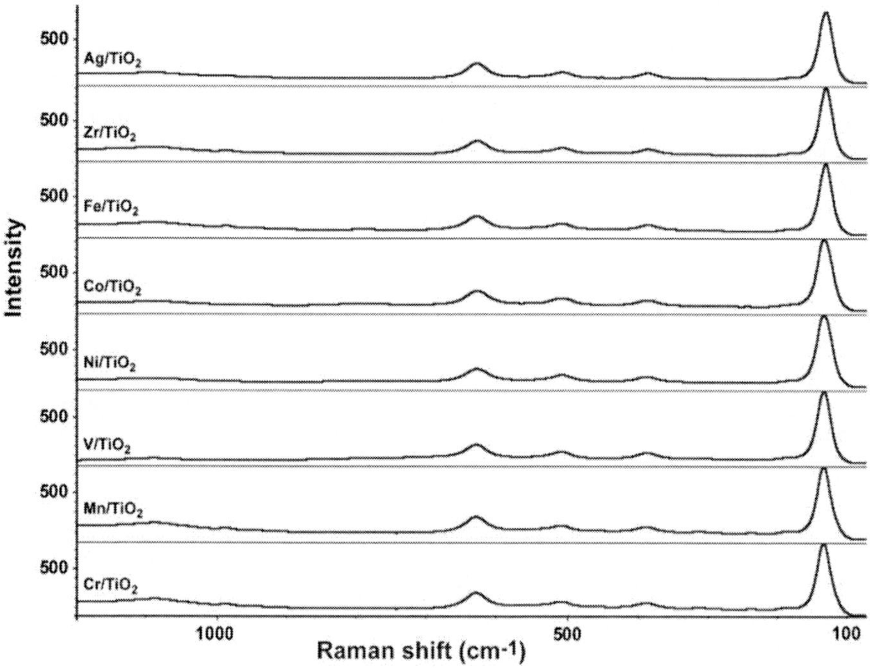
Intensity
500
Ag/TiO2
500
Zr/TiO2
500
Fe/TiO2
500
Co/TiO2
500
Ni/TiO2
500
V/TiO2
500
Mn/TiO2
500
Cr/TiO2
1000
500
100
Raman shift (cm-1)

b.

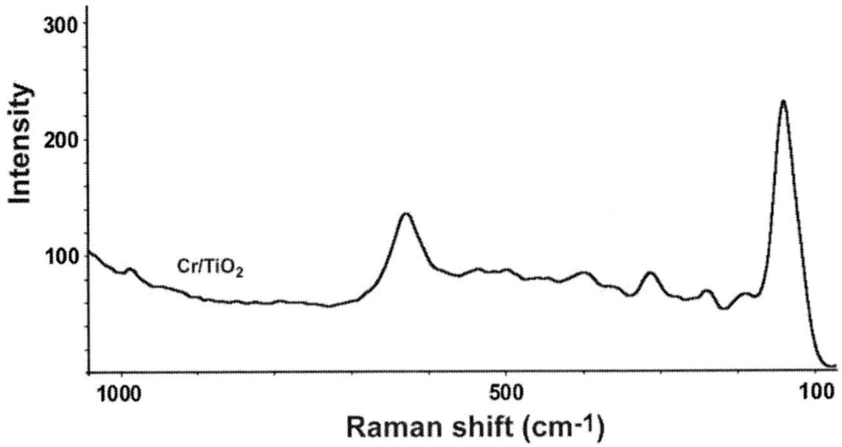

c.

Figure 9. Raman spectra of pure TiO_2 (a) and M/TiO_2 (b-c) (M = Mn, V, Fe, Ag, Ni, Co, Zr, and Cr).

The A_{1g}, B_{1g}, B_{2g} and B_{3g} modes are Raman active, the B_{1u}, B_{2u} and B_{3u} modes are infrared active and the A_{1u} mode is inactive in both Raman and infrared. In total, 17 bands out of the predicted 36 can be easily indentified, including six A_{1g}, six B_{1g}, four B_{2g} and one B_{3g}. Additional modes are masked by a high level of coincidence and weak band intensities. The Raman spectra of synthetic brookite powder [103] exhibit characteristic bands assigned as A_{1g} (127, 154, 194, 247, 412, 640 cm^{-1}), B_{1g} (133, 159, 215, 320, 415, 502 cm^{-1}), B_{2g} (366, 395, 463, 584 cm^{-1}) and B_{3g} (452 cm^{-1}). The overall spectral profile of brookite is one of an intense low-wavenumber bending mode and grouping of weaker, higher wavenumber bands. In comparison, the vibrational spectra of rutile and anatase are relatively simple in accord with their greater crystal symmetry. Figure 9-c shows Raman spectrum of Cr/TiO_2 film which exhibits some characteristic modes, expected for the brookite form of TiO_2: A_{1g} (190, 241, 405, 630 cm^{-1}), B_{1g} (314 cm^{-1}), B_{2g} (360, 395, 498, 585 cm^{-1}) and B_{3g} (445 cm^{-1}).

3.3.3. X-ray Photoelectron Spectroscopy (XPS)

X-ray photoelectron spectroscopy is used to measure the chemical states of surface elements. Valence electrons are more sensitive to changes in the chemical state since these electrons participate in chemical bonding. Higher energy resolution spectra allow users to extract more accurate and detailed chemical shift data from XPS (Figure 10a-k). To analyze the binding energy of the prepared samples, XPS spectra were performed by an ESCA Probe P commercial instrument (Omicron Nanotechnology Ltd, Taunusstein, Germany) using a source of Al Kα-radiation with energy of 1486.7 eV. Measured spectra were analysed by use of CasaXPS programme (Casa Software Ltd., Teignmouth, UK). The binding energies were calibrated with reference to C 1s at 284.7 eV (Figure 10a) and this peak is due to the adventitious hydrocarbon from the XPS instrument itself.

The XPS analyses were made for all prepared samples of doped TiO_2. In this study, the actual binding energy positions corresponding to Ti $2p_{1/2}$ and Ti $2p_{3/2}$ lines for TiO_2 films were found at 465.1 eV and 458.0 eV (Figure 10b). These values agree well with reported XPS data and are known to be due to Ti^{4+} in pure anatase titania form [104]. The binding energy of Zr $3d_{5/2}$ [105] (Figure 10c), Ni $2p_{3/2}$ [106] (Figure 10e), and Fe $2p_{3/2}$ [107] (Figure 10f) lines (181.9, 854.8, and 711.0 eV, respectively) can be assigned to Zr^{4+}, Ni^{2+}, and Fe^{3+}. Owing to the small binding energy difference (1.0 eV for 781.1 eV) and the band broadening caused by multiplet splitting, the presence of Co^{2+} or Co^{3+}, or both, cannot be established from the position of the Co $2p_{3/2}$ and $2p_{1/2}$ lines alone [108] (Figure 10i). The core level of Mn $2p_{3/2}$ and $2p_{1/2}$ shows the binding energy lines at 640.9 and 653.3 eV, respectively (Figure 10g). These peaks with energy separation of 12.4 eV can be assigned to Mn^{2+} [109]. The small peak of V $2p_{3/2}$ (516.0 eV) appeared as two shoulders (Figure 10d). The shoulder at a binding energy of 516.9 eV suggests V^{5+} species, whereas the shoulder at 516.3 eV can be assigned to V^{4+} [110]. Obviously, some V^{4+} ions were oxidized into V^{5+} in the preparation process, because vanadium existed only as V^{4+} in the precursor (as VO^{2+} vanadyl). The formation of V^{5+} possibly occurred during annealing. In the case of Cr/TiO_2 samples, the XPS shows (Figure 10h) two different Cr species which would be to assigned the peak ($2p_{3/2}$) at 576.8 eV, corresponding to Cr^{3+}, and the highest peak at 580.0 eV, corresponding to Cr^{6+} [111,112]. The binding energy of Ag $3d_{5/2}$ peak (367.3 eV) indicated that silver is of metallic nature Ag^0 (Figure 10j) [113]. For the Ag^+ ions spreading on the surface of titania under the action of heat they would be reduced into Ag^0 due to the high redox potential for the Ag^+ ion. Also the Ag-O bonding is much weaker than Ti-O (or Ag-Ag) bonding and Ag has a higher surface free energy than TiO_2 [114,115].

The atomic percentage (at.%) of metals (for 3% w/w metal acac/TiO_2) ranging from 0.12 to 1.52 and the cation valency were obtained for the following metals: Ag^0 (0.12), Fe^{3+} (0.35), Cr^{3+} (0.56), Mn^{2+} (0.99), Ni^{2+} (1.07), V^{4+} and V^{5+} (1.17), Co^{2+} and Co^{3+} (1.20), and Zr^{4+} (1.52).

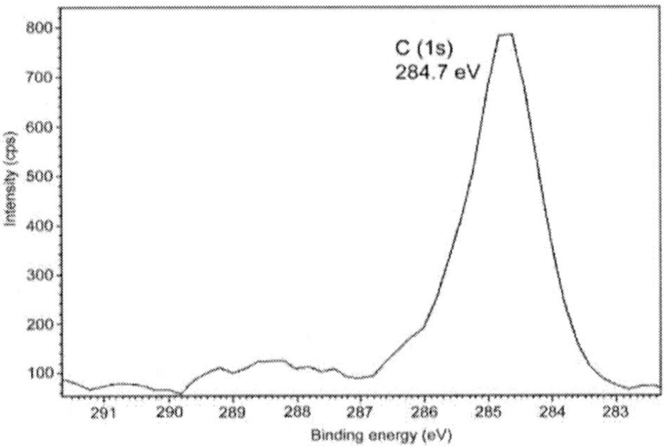
C (1s)
284.7 eV
Intensity (cps)
Binding energy (eV)

a.

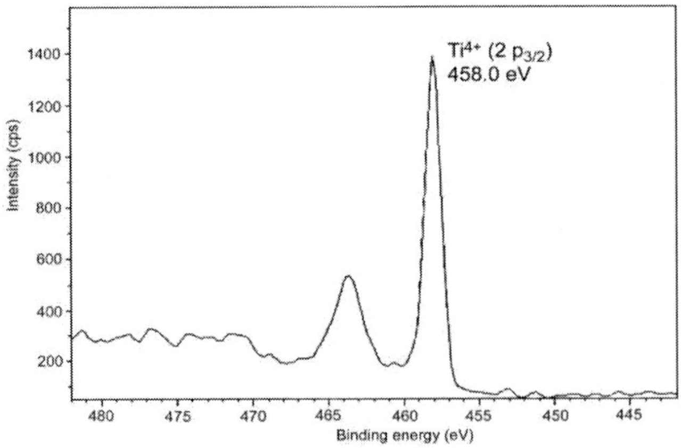
Ti4+ (2 p3/2)
458.0 eV
Intensity (cps)
Binding energy (eV)

b.

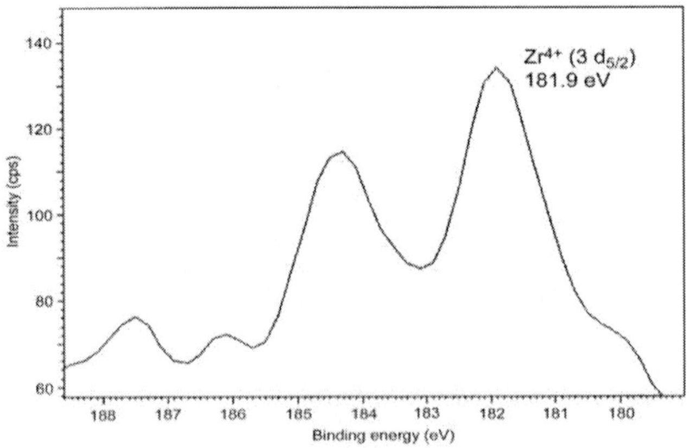
Zr4+ (3 d5/2)
181.9 eV
Intensity (cps)
Binding energy (eV)

c.

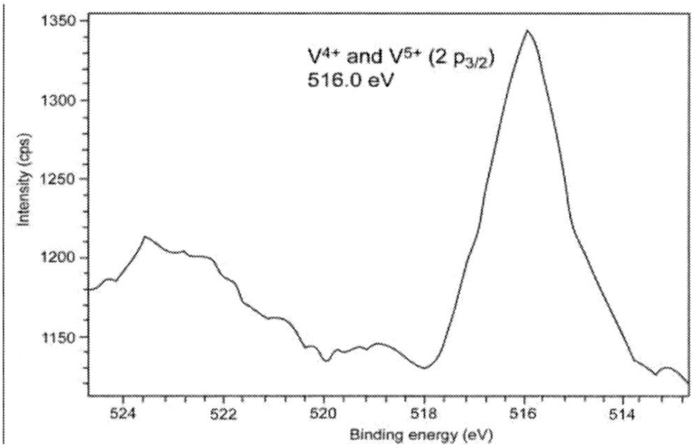
V4+ and V5+ (2 p3/2)
516.0 eV
Intensity (cps)
Binding energy (eV)

d.

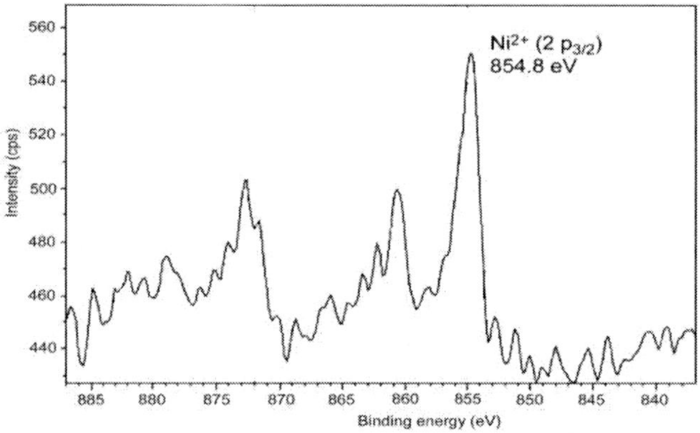
Ni2+ (2 p3/2)
854.8 eV
Intensity (cps)
Binding energy (eV)

e.

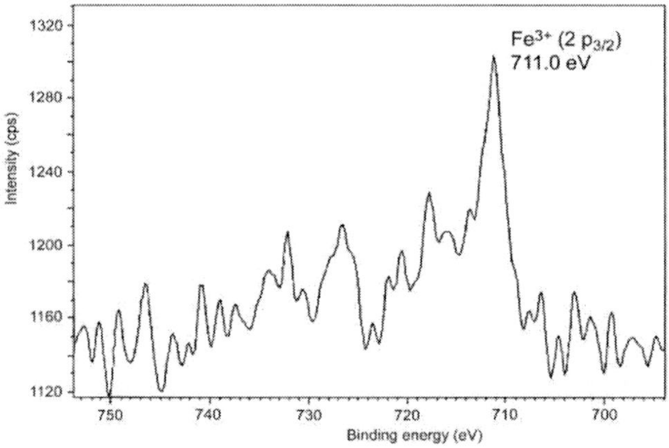
Fe3+ (2 p3/2)
711.0 eV
Intensity (cps)
Binding energy (eV)

f.

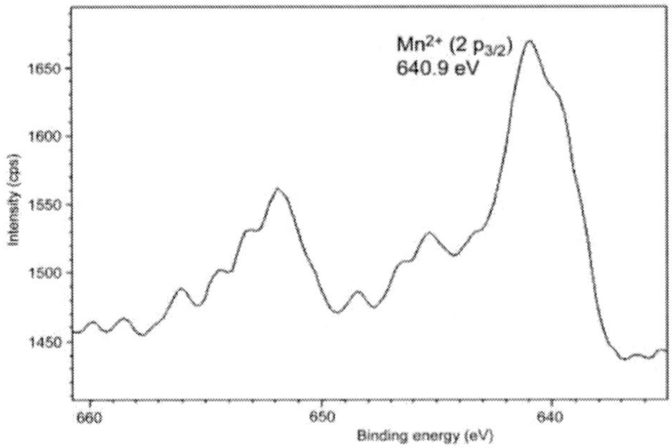
Mn2+ (2 p3/2)
640.9 eV
Intensity (cps)
1650
1600
1550
1500
1450
660
650
640
Binding energy (eV)

g.

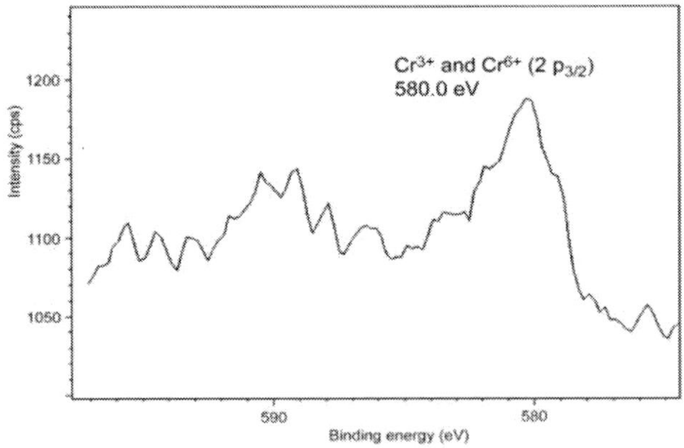
Cr3+ and Cr6+ (2 p3/2)
580.0 eV
Intensity (cps)
1200
1150
1100
1050
590
580
Binding energy (eV)

h.

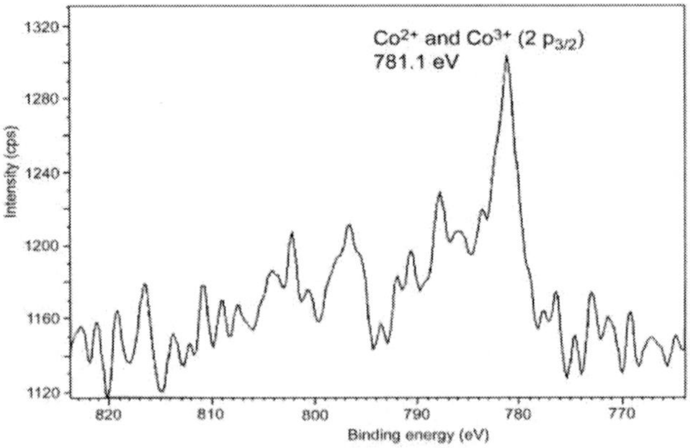
Co2+ and Co3+ (2 p3/2)
781.1 eV
Intensity (cps)
1320
1280
1240
1200
1160
1120
820
810
800
790
780
770
Binding energy (eV)

i.

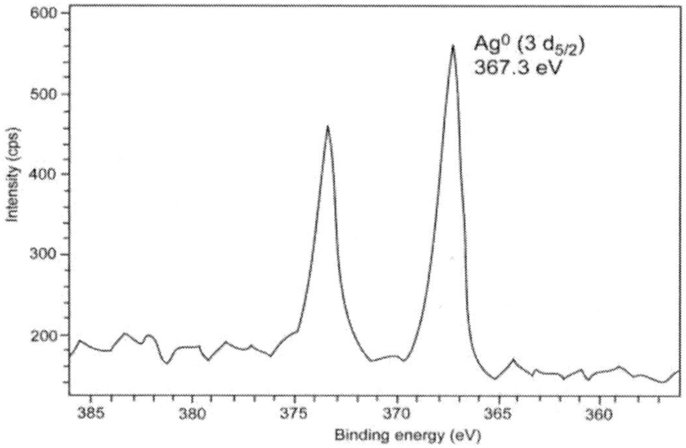

j.

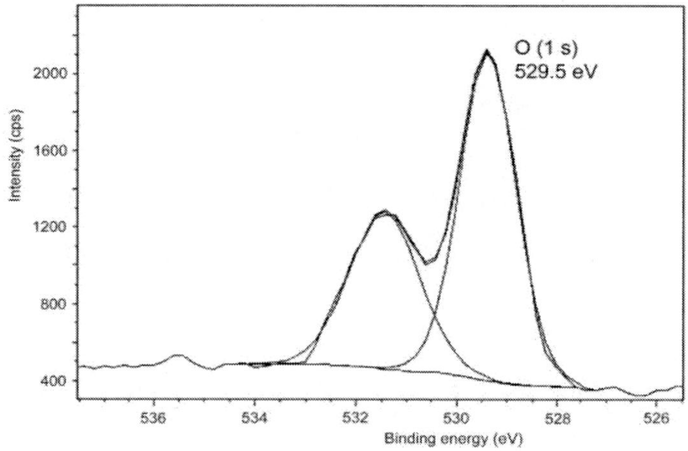

k.

Figure 10. XPS spectra of (a) C, (b) Ti, (c) Zr, (d) V, (e) Ni, (f) Fe, (g) Mn, (h) Cr, (i) Co, (j) Ag, (k) O.

The photoelectron spectra of O 1s (Figure 10k) can be decomposed into two peaks. The O 1s main peak at 529.5 eV is assigned to the metallic oxide, which is consistent with the binding energy of O^{2-} in the TiO_2 lattices. A shoulder to the main O 1s peak can be observed at high binding energy (531.5 eV), which can be attributed to the hydroxyl groups or chemisorbed water molecules adsorbed on TiO_2 surface [116].

3.3.4. Scanning Electron Microscopy (SEM)

The layers morphologies and grain sizes were imaged using scanning electron microscopy. The micrographs were recorded with FEI (Quanta 200; FEI Co., Hillsboro, OR, USA) at 10 kV. The samples for SEM imaging were mounted on metal stubs with a piece of conducting tape then coated with a thin layer of gold film to avoid charging. SEM images of titanium dioxide layer are shown in Figure 11a. The average layer thickness is about 130 nm (Figure 11b). The images reveal smooth morphologies and cluster-free surfaces. Similar morphologies were obtained for the zirconium, vanadium and nickel doped TiO_2 films (Figure 12a-c).

a.

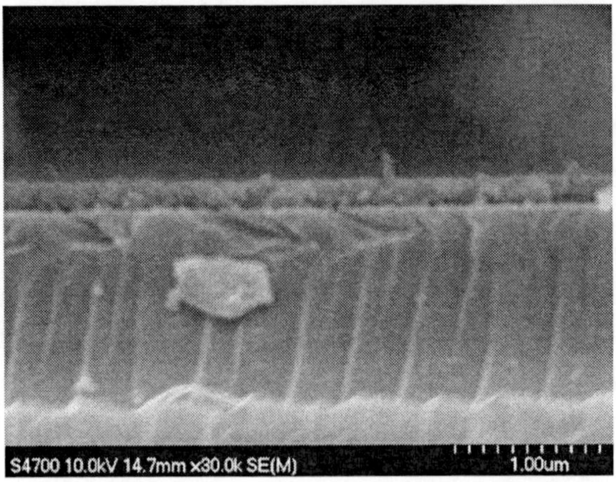

b.

Figure 11. SEM micrographs of TiO_2 thin film: (a) surface, (b) cross-section.

S4700 10.0kV 12.7mm x1.50k SE(M)
30.0um

a.

S4700 10.0kV 14.2mm x3.00k SE(M)
10.0um

b.

S4700 10.0kV 13.0mm x1.50k SE(M)
30.0um

c.

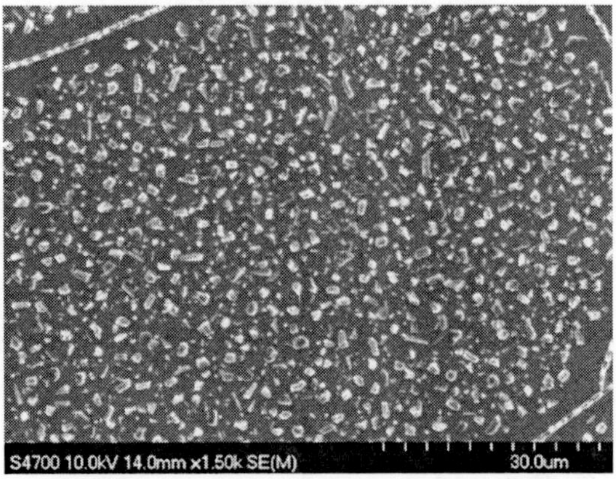
S4700 10.0kV 14.0mm x1.50k SE(M)
30.0um

d.

S4700 10.0kV 13.5mm x1.50k SE(M)
30.0um

e.

S4700 10.0kV 13.7mm x1.50k SE(M)
30.0um

f.

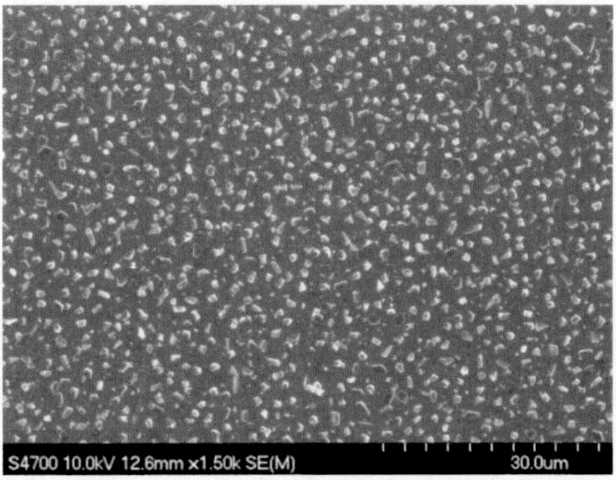

g.

h.

Figure 12. SEM micrographs of titania-doped film: (a) Zr, (b) V, (c) Ni, (d) Fe, (e) Ag, (f) Mn, (g) Co, (h) Cr.

Micrographs of other titania-doped layers are shown in Figure 12d-h. In general, the particles are agglomerated (10-30 nm) and basically irregular in shape, with a substantial variation in particle size. Nevertheless, the SEM micrographs showed that Fe/TiO_2, Ag/TiO_2, Mn/TiO_2, Co/TiO_2 and Cr/TiO_2 layers aggregated to bigger particles. The aggregation of Ag/TiO_2 is less serious and there is no aggregation of Zr/TiO_2, V/TiO_2 and Ni/TiO_2 was observed and the surface morphology of undoped TiO_2 is much smoother than the others. It is interesting to note, that Ag/TiO_2 film exhibited different surface morphology than the other titania-doped films (Figure 12e). It has been reported that a small addition of Ag produces nucleation sites that can induce the formation of aggregates or coalescence during processing [117].

3.3.5. Atomic Force Microscopy (AFM)

Surface pictures in 3D were taken on the AFM Explorer (ThermoMicroscopes, Sunnyvale, USA) and these were then analyzed to estimate the parameter of the relative surface roughness. Parameter rms (root mean square) comes from the fact that it is the square root of the mean of the squares of the values. The AFM image can provide some quantitative data about the surface roughness and also grain size of the TiO_2 thin film. AFM images of TiO_2 and doped TiO_2 films have been shown in Figure 13a-i. Analysis of the microstructure performed with AFM, shows that the thin films consisted of spherical grains of 10 nm - 40 nm in the diameter. The surface of the films appears uniform and homogeneous. The root mean square roughness value of the pure titania thin film surface is 0.48 nm. AFM images of titania-doped thin films exhibited granular nanostructured morphology. The primary particles to construct thin films of TiO_2 are observed to be smaller than those of the doped TiO_2 thin films.

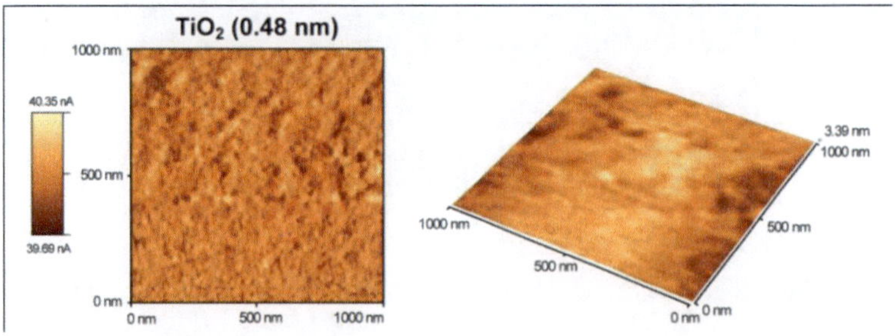

a.

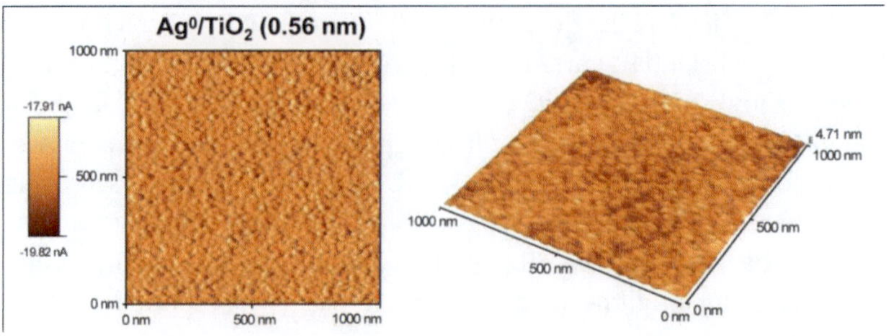

b.

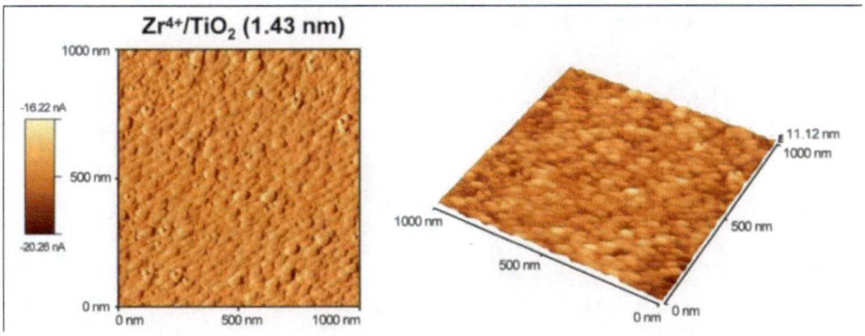
Zr4+/TiO2 (1.43 nm)
1000 nm
-16.22 nA
500 nm
-20.26 nA
0 nm
0 nm
500 nm
1000 nm
11.12 nm
1000 nm
1000 nm
500 nm
500 nm
0 nm
0 nm

c.

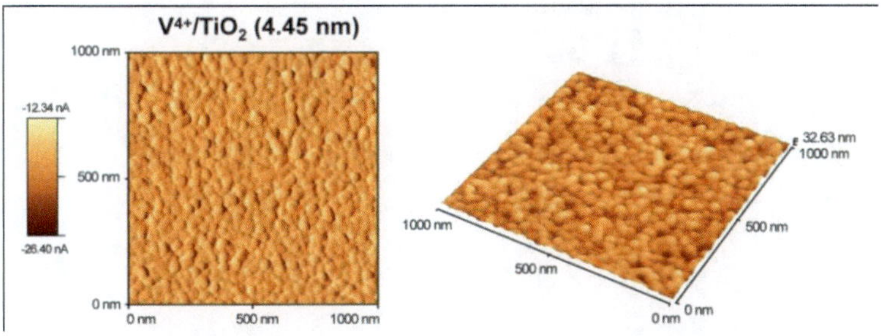
V4+/TiO2 (4.45 nm)
1000 nm
-12.34 nA
500 nm
-26.40 nA
0 nm
0 nm
500 nm
1000 nm
32.63 nm
1000 nm
1000 nm
500 nm
500 nm
0 nm
0 nm

d.

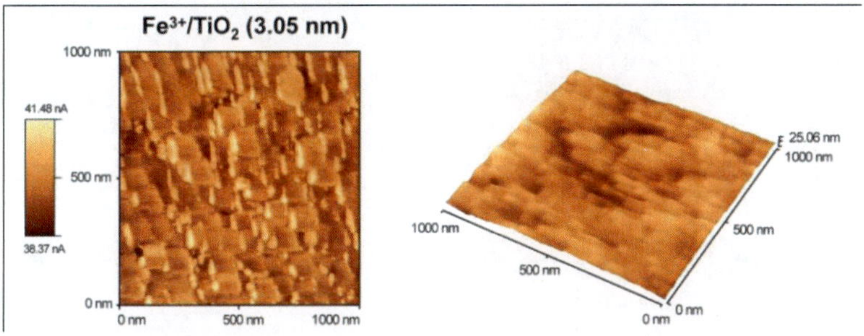
Fe3+/TiO2 (3.05 nm)
1000 nm
41.48 nA
500 nm
38.37 nA
0 nm
0 nm
500 nm
1000 nm
25.06 nm
1000 nm
1000 nm
500 nm
500 nm
0 nm
0 nm

e.

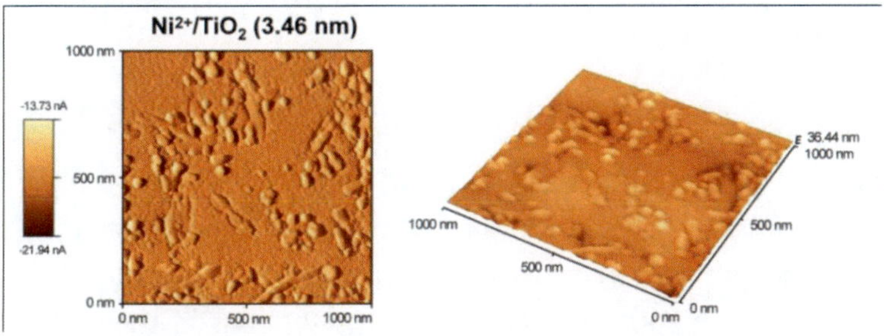
Ni2+/TiO2 (3.46 nm)
1000 nm
-13.73 nA
500 nm
-21.94 nA
0 nm
0 nm
500 nm
1000 nm
36.44 nm
1000 nm
1000 nm
500 nm
500 nm
0 nm
0 nm

f.

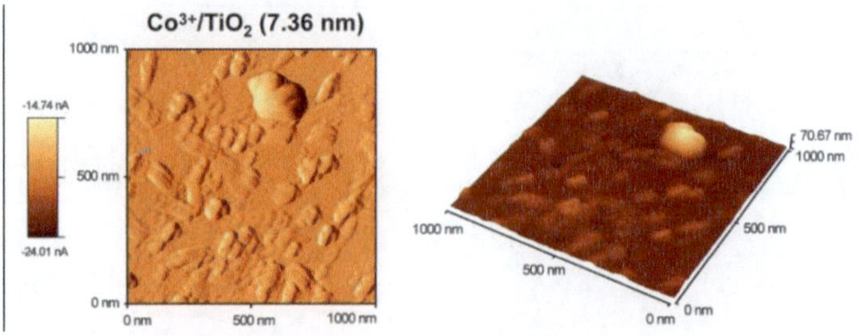

g.

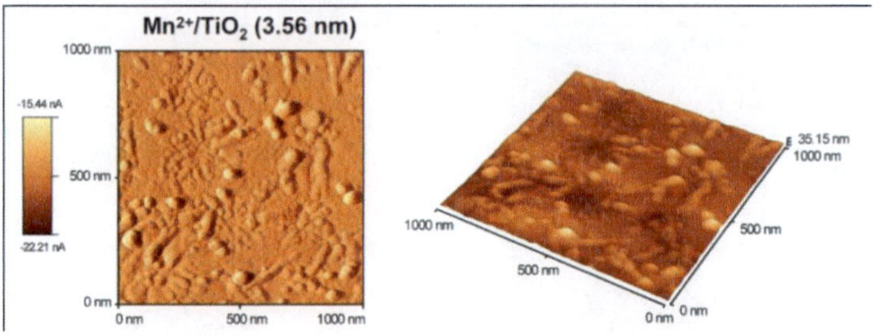

h.

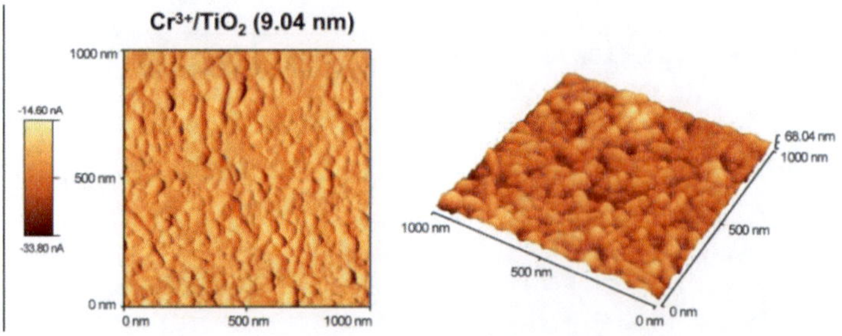

i.

Figure 13. AFM images of titania-doped thin films (3% w/w metal acac/TiO_2): (a) pure TiO_2, (b) Ag, (c) Zr, (d) V, (e) Fe, (f) Ni, (g) Co, (h) Mn, (i) Cr (in parentheses is relative surface roughness).

3.3.6. UV/Vis Absorption Spectroscopy

The absorption edges were determined using absorption data of thin films recorded by UV/Vis spectrophotometer (Shimadzu, UV-2450, Japan) in the range of 200-800 nm.

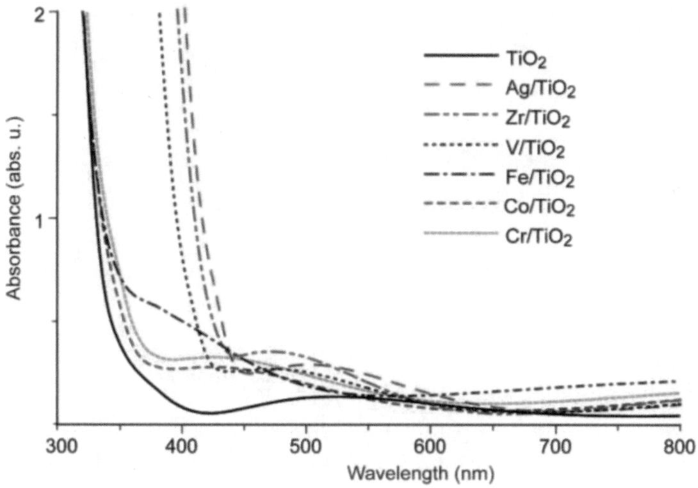

Figure 14. UV/Vis absorption spectra of selected titania-doped thin films.

UV/Vis spectra clearly show that the incorporation of Ag, Zr, and V ions into TiO_2 leads to a red shift in the optical response, as well as reduction in the bandgap energy (Figure 14). The largest shifts have occurred for Ag/TiO_2 (424 nm), Zr/TiO_2 (420 nm) and V/TiO_2 (410 nm). Other doped samples of TiO_2 exhibited absorption only in the UV region: Co/TiO_2 (368 nm), Mn/TiO_2 (369 nm), Cr/TiO_2 (372 nm), Ni/TiO_2 (386 nm), Fe/TiO_2 (405 nm). Just absorption spectrum of Fe/TiO_2 showed a very weak red shift. The red shift in absorption spectra and reduction in bandgaps are attributed to the transfer of electrons from transition metal ions to the conduction band of TiO_2. The absorption edge of undoped titanium dioxide thin film was determined as 365 nm, which is in good agreement with literature.

3.4. Photocatalytic Activity of Thin Titania Films

The photocatalytic decomposition of organic pollutants in water generally follows the Langmuir-Hinshelwood mechanism [118]:

$$r = \frac{kKc}{1 + Kc} \tag{2}$$

where r is the photocatalytic reaction rate (mol l^{-1} min^{-1}), k is photocatalysis rate constant (mol l^{-1} min^{-1}), K is equilibrium adsorption coefficient (l mol^{-1}) and c is reactant concentration (mol l^{-1}) at time t. At low substrate concentrations, the term Kc (Eq. 2) in the denominator can be neglected with respect to unity and the reaction rate becomes the apparent first order, i.e. $r = kKc$. This kinetic equation may be written in the integral form:

$$\ln\left(\frac{c_0}{c}\right) = kKt = k_{app}t \tag{3}$$

where c_0 is the initial reactant concentration (mol l^{-1}) and k_{app} is the apparent photocatalysis rate constant (min^{-1}) of the pseudo-first order reaction. For convenient comparison of the photocatalytic activity of the different thin films, k_{app} has been chosen as the appropriate kinetic parameter, since it enables the determination of photocatalytic activity independent of the previous adsorption period in the dark [119].

The photocatalytic activity of titania films was examined by the decomposition test of Rhodamine B (RhB). The degradation process, involving hydroxyl radical formation and subsequent degradation of RhB by the OH·, obeys apparent first-order kinetics [120]. The apparent rate constants of RhB (k_{app}), obtained from the slopes (Eq. 3), and conversions (x) of RhB, are reported in Table 1.

The TiO_2 films, pure and also doped, as prepared have effectively decomposed the RhB and the conversions were 19-61 % (Table 1) after 90 min irradiation. It can be seen that after 90 min irradiation the RhB conversion on the pure TiO_2 film was not more than 41 % and the apparent rate constant was 0.0074 min^{-1}. For TiO_2 films doped with Ag, Zr, and V, the conversion was enhanced significantly from 46 to 61 %. The best apparent degradation rate constant (0.0125 min^{-1}), which was higher than that on the pure TiO_2 film by 1.7 times, was obtained on the Ag(3%)/TiO_2 photocatalyst.

To compare the doping effect of Ag, Zr, and V acetylacetonate (w/w% metal/TiO_2) on the photocatalytic activity, the amount affording the maximum conversion was determined in all cases. The Ag(3%)/TiO_2, Zr(3%)/TiO_2, and V(5%)/TiO_2 samples showed the highest values (Table 1).

The photodecomposition rate could be enhanced because more charge carriers are available. According to the principle of photocatalysis, in the V, Zr, and Ag-doped TiO_2 films metal can provide a shallow trap for the photogenerated electron and hole inhibiting recombination and extending the lifetime of the charge carrier.

Chao [121] found that Ag-doping promotes the anatase to rutile transformation, which is attributed to the increase in the specific surface area and enhances the electron-hole pair separation.

Some dopants can affect negatively the photoreactivity of TiO_2 (Table 1; Fe, Ni, Co, Mn, and Cr), changing the number of active sites, the surface group type or the acid-base properties [65]. The different behavior of the various samples is also related to the solubility of the transition metal in the support, which depends on the ion radius and charge [65]. In this study, chromium was found to be the worst dopant among the various metals investigated (Table 1, conversion 21 %). It has low diffusion length and also serves as a trap for the electrons produced under irradiation.

Table 1. Photodecomposition of Rhodamine B (RhB) with the initial concentration 10^{-4} mol l^{-1} in 90 min on titania-doped thin films (w/w % metal acac/TiO_2)

Catalyst	RhB apparent rate constant k_{app} (min^{-1})	Conversion (%)
TiO_2	0.0074	41
Ag(1%)/TiO_2	0.0087	54
Ag(3%)/TiO_2	0.0125	61
Ag(5%)/TiO_2	0.0085	53
Zr(1%)/TiO_2	0.0097	58
Zr(3%)/TiO_2	0.0119	59
Zr(5%)/TiO_2	0.0079	46
V(1%)/TiO_2	0.0047	34
V(3%)/TiO_2	0.0076	42
V(5%)/TiO_2	0.0079	46
V(9%)/TiO_2	0.0024	19
Fe(3%)/TiO_2	0.0061	36
Ni(3%)/TiO_2	0.0059	34
Co(3%)/TiO_2	0.0053	31
Mn(3%)/TiO_2	0.0042	25
Cr(3%)/TiO_2	0.0038	21

Chapter 4

INTERACTIONS OF MW RADIATION WITH THE UV/VIS-ILLUMINATED TITANIA

Although *microwave chemistry* has already received widespread attention from a chemical community, considerably less information is available about the effect of MW radiation on photocatalytic reactions. The MW region of the electromagnetic spectrum, on the other hand, lies between infrared radiation and radio frequencies. Its energy (E = 1–100 J mol^{-1} at ν = 1–100 GHz) is approximately 3–6 orders of magnitude lower than that of UV/Vis radiation. MW heating is not identical with classical external heating, at least at the molecular level. Molecules with a permanent (or induced) dipole respond to an electromagnetic field by rotating, which results in friction with neighboring molecules (thus generating heat). Additional (secondary) effects of microwaves include ionic conduction (ionic migration in the presence of an electric field) or spin alignment [13].

MW radiation provides not only the heat source (as thermal effect) but also a specific effect [122,123] (non-thermal effect), as suggested by Marken et al. [124], that leads to the enhanced photo-assisted degradation of several substrates. Booske et al. [125] have suggested that microwaves can couple into the low MW frequency elastic lattice oscillations of a crystalline solid, thereby generating a non-thermal distribution that might enhance the ion mobility and thus the diffusion of charge carriers to the surface leading to the increased formation of the surface •OH radicals and to the increased concentration of electrons at the surface [126].

Specific interactions of microwaves with the UV-illuminated TiO_2 particle surface may give rise to the generation of additional surface defects [127] that can directly increase the concentration of •OH radicals, or some other reactive oxygen species, in the aqueous dispersion [126]. The formation of •OH radicals was probed by the ESR technique, in which the sample could be irradiated by both UV light and MW radiation. In this regard, EPR spectroscopic evidence showed that a greater number of •OH radicals were produced in the MW-assisted photocatalytic process in aqueous TiO_2 dispersions [15]. ESR spectra and the photooxidation experiments confirmed that the mixed-phase (P-25) TiO_2 specimen was photocatalytically more active than either of the pure anatase or rutile phases of titania [128]. Horihoshi et al. [14] proved by ESR that about 20 % more •OH radicals were generated by the photocatalysis with MW irradiation than by photocatalysis alone.

Blount et al. [129] showed that TiO_2 thin films were more active for photocatalysis of acetaldehyde, acetic acid, and toluene than TiO_2 in the powder form. Also Martyanov et al. [130] described higher activity of prepared sol-gel layers than of the TiO_2 powder.

Chapter 5

NOVEL MICROWAVE PHOTOCATALYTIC REACTORS

While some of the physico-chemical principles of photocatalysis are relatively well understood [77], the reactor engineering of photocatalytic units still requires consideration. Photocatalytic reactor design presents unique challenges due to the requirement that a high surface-area particulate catalyst, a UV/Vis illumination source, and contaminant species be kept in close proximity.

The photocatalytic reactor utilized for MW-assisted experiments is an essential tool for experimental work. Such an equipment enables simultaneous irradiation of the sample with both MW and UV/Vis radiation. Typical reactor designs fall into two general classes: batch and continuous-flow.

5.1. BATCH EXPERIMENTAL SET-UP

Církva and Hájek have proposed a simple application of a domestic MW oven for MW-assisted photochemistry experiments [12]. In this arrangement, EDL was placed in a reaction vessel located in the cavity of an oven. The MW field generated a UV/Vis discharge inside the lamp that resulted in simultaneous UV/Vis and MW irradiation of the sample. This arrangement provided the unique possibility of studying photochemical and photocatalytic reactions under extreme thermal conditions [131].

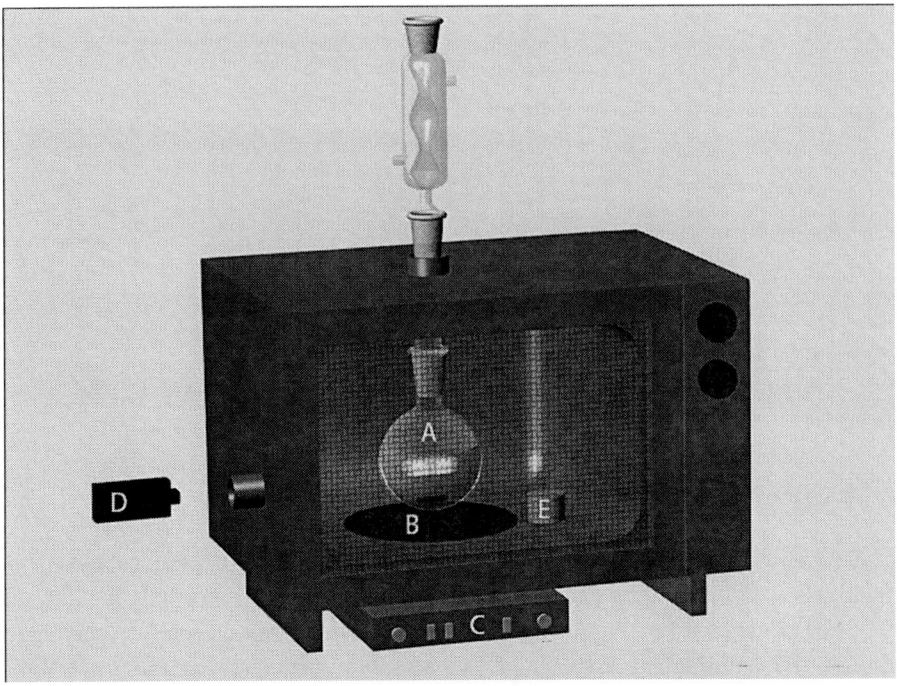

Figure 15. Modified MW oven for MW-assisted photochemical or photocatalytic experiments: (A) reaction mixture with EDL and a magnetic stir bar; (B) aluminum plate; (C) magnetic stirrer; (D) fiber optic spectral probe; (E) dummy load inside the oven cavity [132].

Klán, Literák et al. published a series of papers that described the scope and limitation of this reactor [132-135]. In a typical design (Figure 15), two holes were drilled into the walls of a domestic oven, one for a condenser tube in the oven top and second in the side for a fiber optic spectral probe. Part of the oven bottom was replaced with an aluminum plate to enable magnetic stirring. The certain amount of a MW-absorbing solid material (dummy load - basic Al_2O_3, molecular sieve, etc.) was inserted when a small quantity of a non- or poorly absorbing sample was used. The material removed excess MW power and prevented the magnetron from being destroyed by overheating. The experimental equipment consists from a round-bottom flask (500 ml) equipped with the Dimroth condenser. For reactions carried out in air, the required air (0.03 $m^3.h^{-1}$) was bubbled through a porous glass tube sparger located at the bottom of the reactor. EDL had always to be placed in a position, in which the solvent cooled it efficiently, because lamp overheating might cause failure of the lamp emission. Intense IR output from the lamp triggered immediate boiling of all solvents including nonpolar (MW-transparent) liquids [133,134]. Polar solvents (water for photocatalysis), on the other hand, absorbed most of MW radiation, resulting in reduction UV/Vis output efficiency. In Table 2 are depicted the most important advantages and disadvantages of EDL applications. The laboratory MW oven (MicroSYNTH Labstation, 1000 W; Milestone, Italy) for photocatalytic experiments is shown in Figure 16.

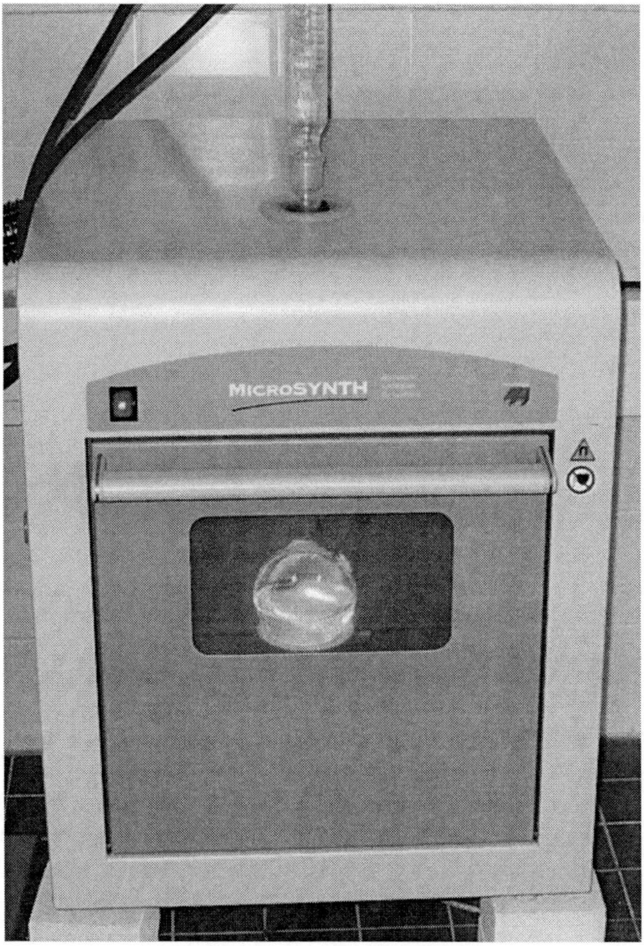

Figure 16. Microwave photocatalysis arrangement on Milestone's MicroSYNTH Labstation.

Table 2. Advantages and disadvantages of EDL applications in photocatalysis

Advantages
• Simultaneous UV/Vis and MW irradiation of the sample
• Simplicity of the experimental set-up (use of a commercially MW oven, "wireless" EDL operation)
• Low cost of EDL (easy method of EDL preparation in lab)
• Choice of the EDL material (Hg, S) might modify its spectral output
• Good photocatalytic efficiency – EDL is "inside" the sample
• Possibility of performing photocatalysis at high temperature
Disadvantages
• Technical difficulties of performing photo-experiments at temperatures below the solvent b.p.
• Intensity of EDLs strongly depends on given experimental conditions: a) b.p., polarity, and transmittance of solvent b) output and sort of MW equipment c) type and intensity of cooling

Microwave-assisted photocatalysis in a batch can be facilitated by a combination of the following three variables: the type and variation of semiconductor photocatalyst (slurry vs. thin film), external or internal lamp (classical UV lamp vs. electrodeless discharge lamp, EDL), and the frequency of the microwave field (2.45 GHz). The combination of the first two variables (photocatalyst and lamp) may lead to the following four types of techniques:

a) TiO_2 (slurry) + classical UV lamp [14,15,126,128,136-140];
b) TiO_2 (slurry) + EDL [122,123,141-146];
c) TiO_2 (thin film) + classical UV lamp [147];
d) TiO_2 (thin film) + EDL [148].

Several studies have been reported in the last few years on the enhanced efficiency of the photocatalytic degradation of organic substrates using a combination of classic UV light and MW radiation in the treatment of aqueous TiO_2 dispersions (*type a*) of humic acid [136], rhodamine-B dye [14, 137], 2,4-dichlorophenoxyacetic acid (2,4-D) [126], phenol [128], bisphenol-A [138, 139], 4-chlorophenol [15], and other compounds [140].

Recent papers have demonstrated outstanding improvement of the degradative efficiency by the coupling of MW radiation with electrodeless discharge lamps (EDLs) (*type b*) to the photocatalytic degradation of various substrates, e.g. rhodamine-B [141], 2,4-dichlorophenoxyacetic acid (2,4-D) [122,123], phenol [142], 4-chlorophenol [143], azo dye brilliant red X-3B [144], and methylene blue [145]. Some high-temperature and high-pressure microwave/UV digestion procedure [146] of biological samples (serum, urine, milk) in the aqueous TiO_2 suspension has been accomplished using a cadmium low-pressure EDL. Comparative studies showed that EDL may provide a high yield of the product in shorter reaction time [12].

As an example (*type c*) of the combination of classic UV light and MW radiation on the thin film layer of the photocatalyst (TiO_2/ZrO_2 on glass beads) the gaseous phase photocatalytic oxidation of ethylene has been performed [147] only. The photocatalytic reactions in a solution on some thin film layer of the photocatalyst (*type d*) have been published by us [148] (see chapter 6).

5.2. Continuous-Flow Experimental Set-up

Církva and coworkers [149] have studied a continuous-flow MW photoreactor, which contained the glass tube with quartz Hg-EDL (a 254 nm emission) inside a MW oven. Photohydrolysis of mono-chloroacetic acid to hydroxyacetic acid and HCl has been chosen as a model reaction. The course of the reaction was followed by a pH change of the solution. The conversion was optimized as a result of a trade-off between the thermal dependence of quantum yield (which increases with increasing temperature) and the thermal dependence of relative intensity of a short-wavelength band (which increases with decreasing temperature).

The experimental set-up for the continuous-flow MW photocatalytic reactor [150] is shown in Figure 17. The aqueous mixture was circulated through the photoreactor system consisting of the glass reservoir (A) (500 mL; fitted with a thermometer (B) and a chloride ion-selective electrode (C)) connected by tubing (ISO Versinic; Saint-Gobain, France) to a PTFE diaphragm pump (E) (Cole-Parmer, Masterflex; Vernon Hills, IL, USA), a glass tube (200 mm (length) x 22 mm (internal diameter)) with titania-coated EDLs inside a modified MW oven (Panasonic, NN-Q543WF, 1000 W, inverter), and cooling condenser (H). When the constant MW power was switched on, the titania-coated EDLs started to emit UV/Vis light (measured by spectrometer with a fiber-optic probe, G). The desired temperature in the continuous-flow photoreactor was achieved by adjustment of the pump flow speed. For reactions carried out in air, the required air (flow 0.03 $m^3\ h^{-1}$) was bubbled in via inlet (C) (porous glass tube) located on the neck of reservoir (A).

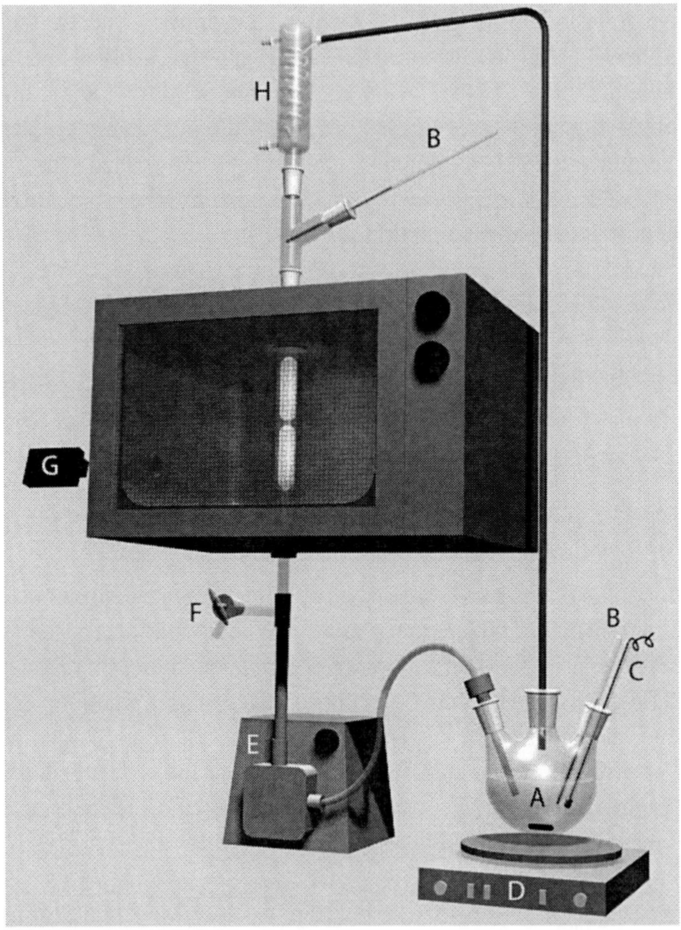

Figure 17. Continuous-flow microwave photocatalytic reactor: (A) glass reservoir with magnetic stir bar; (B) thermometer; (C) chloride ion-selective electrode and/or tube for air bubbling; (D) magnetic stirrer; (E) PTFE diaphragm pump; (F) outlet; (G) spectrometer with a fiber-optic probe; (H) cooling condenser.

Microwave photocatalysis in continuous-flow reactors can be facilitated by a combination of the following two variables: the type and variation of semiconductor photocatalyst (slurry vs thin film), and external or internal lamp (classical UV lamp vs electrodeless discharge lamp; EDL). The combination of variables may lead to the following four types of techniques for continuous-flow set-ups in the microwave field:

a) TiO_2 (slurry) + classical UV lamp [151];

b) TiO_2 (slurry) + EDL [126,141,143,144,151-155];
c) TiO_2 (thin film) + classical UV lamp [147];
d) TiO_2 (thin film) + EDL [150,156].

The enhanced efficiency of photocatalysis in a continuous-flow microwave reactor using a combination of classical UV light and MW radiation has been reported in the treatment of aqueous TiO_2 dispersions (type a) of Rhodamine-B dye [151].

Recent publications have reported significant advancement in the degradative efficiency of continuous-flow reactors through the coupling of EDLs with MW radiation (type b) to generate UV/Vis irradiation resulting in photocatalytic degradation or decomposition of various substrates, e.g. Rhodamine-B [141,151], 2,4-dichlorophenoxyacetic acid (2,4-D) [126], 4-chlorophenol [143,152-154], and azo dye brilliant red X-3B [144,155].

As an illustration (type c) of the combination of classic UV light and MW radiation in continuous-flow reactors on the thin film layer of the photocatalyst (TiO_2/ZrO_2 on glass beads) only the gaseous phase photocatalytic oxidation of ethylene has been performed [147].

Photocatalytic reactions in the flowing solution on thin film layers of photocatalyst (type d) with EDLs have been mentioned only in a patent [156] and by us [150].

Chapter 6

MICROWAVE PHOTOCATALYSIS WITH TITANIA-COATED ELECTRODELESS LAMPS

Mono-chloroacetic acid (MCAA) was chosen as a model compound for the microwave-assisted photocatalytic reactions on the titania thin films due to the easy determination of chloride ions (by an ion-selective electrode). Several research groups claim that MCAA doesn't undergo direct photolysis [157-159] and the oxidative degradation and complete decomposition of MCAA by heterogeneous photocatalysis has been investigated [160-164]. The complete MCAA decomposition by oxidation can be written as follows:

$$2\ ClCH_2COOH + 3\ O_2 \rightarrow 4\ CO_2 + 2\ H_2O + 2\ HCl$$

In each experiment, the photocatalytic reactor was filled with 150 ml (batch type, Figure 15), or 500 mL (continuous-flow type, Figure 17) of an aqueous solution of mono-chloroacetic acid (0.1 $mol.l^{-1}$). Then the titania coated mercury EDL was placed into the reaction mixture, or a reactor tube, and MW field was switched on to generate UV/Vis radiation. Samples were taken with a 5 ml syringe from the reactor after every 20 minutes and immediately analyzed by a chloride ion-selective electrode (Chloride Flow Plus Combination ISE, Sentek, Braintree, UK). ISE was connected to a digital pH/mV meter (inoLab Level 1, WTW, Weilheim, Germany) calibrated with NaCl. Results of the ISE analysis were compared with those obtained by the mercurimetry [165] with mercury(II) perchlorate. For mono-chloroacetic acid no additional reaction intermediate as the formaldehyde [164] was detected (2,4-dinitrophenylhydrazine test) under these thermal reaction conditions.

6.1. Effect of Electrodeless Lamps

6.1.1. Effect of the Number of Coating Cycles (in Batch Type)

The effect of the number of coating cycles on the photocatalytic degradation of MCAA was examined. Figure 18 shows that there were no significant differences between numbers of coating cycles on the formation of chloride ions because the reaction takes place on the photocatalyst surface. Generally, it is not possible to prepare a layer of appropriate properties during one coating cycle. Thinner layers are usually difficult to activate, whereas thicker layers don't have some of the important properties, such as high transparency, good mechanical strength, or excellent adhesion to various supports. For these reasons all of the TiO_2 film samples employed herein were coated in four layers.

The complete decomposition of MCAA (0.1 mol.l^{-1}) to HCl, CO_2, and H_2O was accomplished in 490 min (the EDL intensity: 5.56 μW.cm^{-2}; EDL: 2 layers), when the concentration of chloride ions was 0.0997 mol.l^{-1}.

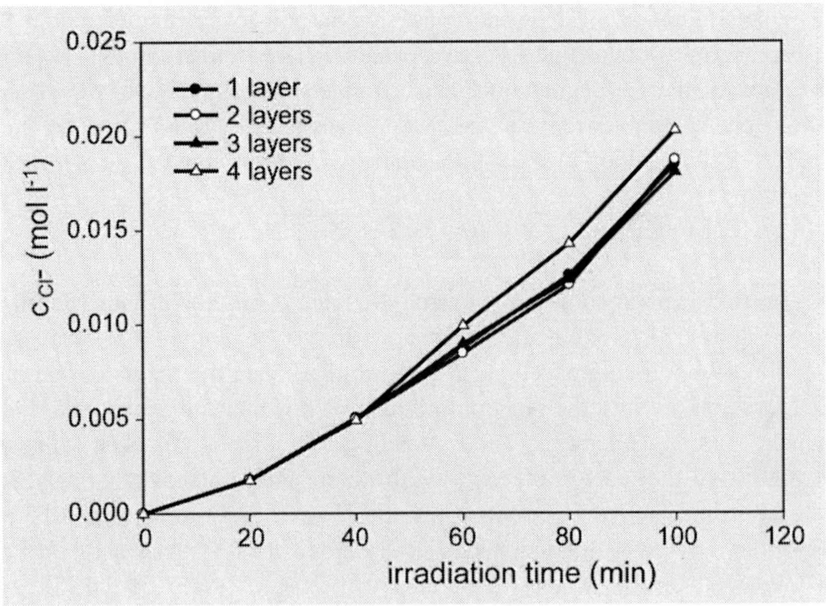

Figure 18. Effect of the number of coating cycles on the concentration of chloride ions in time (EDL intensity: 5.56 μW.cm^{-2}).

6.1.2. Effect of Light Intensity (in Batch Type)

The light intensity effect was investigated in the reaction mixture amounting 150 ml and at the initial MCAA concentration of 0.1 mol l^{-1}. The light intensity was varied from 3.9 $\mu W.cm^{-2}$ to 8.7 $\mu W.cm^{-2}$. Figure 19 shows the change in the chloride ion concentration with light intensity. The photocatalytic reaction efficiency increased with increasing light intensity presumably because more active sites can be formed on the catalyst surface. The influence of light intensity depends on the reaction conditions, such as the reaction mixture amount and, of course, the dissolved pollutant amount. Under lower light intensity an insufficient number of active sites for the degradation of the pollutant in low concentration can be formed. In such cases, the photocatalytic reaction proceeds more slowly and the higher light intensity doesn't have any influence on the photocatalytic efficiency.

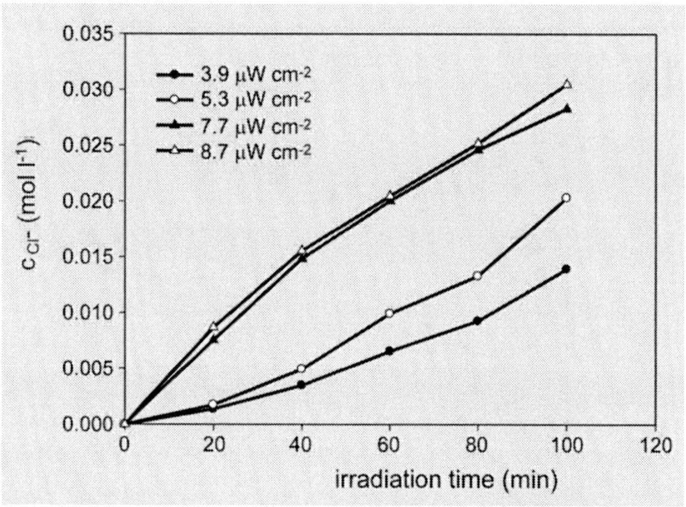

Figure 19. Time dependence of c_{Cl^-} at various intensity of light (EDL: 1 layer).

6.1.3. Effect of Number of the Coated EDLs (in Continuous-flow type)

The number of titania-coated EDLs has an effect on the light intensity, which has been investigated in a batch type photoreactor (see chapter 6.1.2.). The photocatalytic MCAA reaction efficiency increased proportionally with increasing light intensity (radiant flux Φ) of the EDLs [166]. This confirms the photo-induced nature of the activation of the catalytic process, with the participation of photo-induced electrical charges (electrons and holes) in the reaction mechanism [166]. However, above a certain value (25 mW cm^{-2}) [166], the reaction rate becomes proportional to $\Phi^{1/2}$. It can be demonstrated that the rate of electron-hole formation becomes greater than the photocatalytic rate, which favours electron-hole recombination. Another report stated that at high intensities [167], the expected rate-limiting factor is mass transfer, i.e. the rate of photodegradation becomes independent of light intensity.

Figure 20 shows the influence of number of the titania-coated EDLs on MCAA photodegradation at 90 °C. The light intensity of one EDL at 90 °C is only about 2 μW cm^{-2} (for the 366 nm mercury line) which is much less than 25 mW cm^{-2} (for all lines present). An increase in the number of lamps (from 1 to 3 lamps) increased the concentration of chloride ions in time during irradiation. This finding corresponds with literature results where higher light intensity and amount of photocatalyst led to higher degradation rate [160].

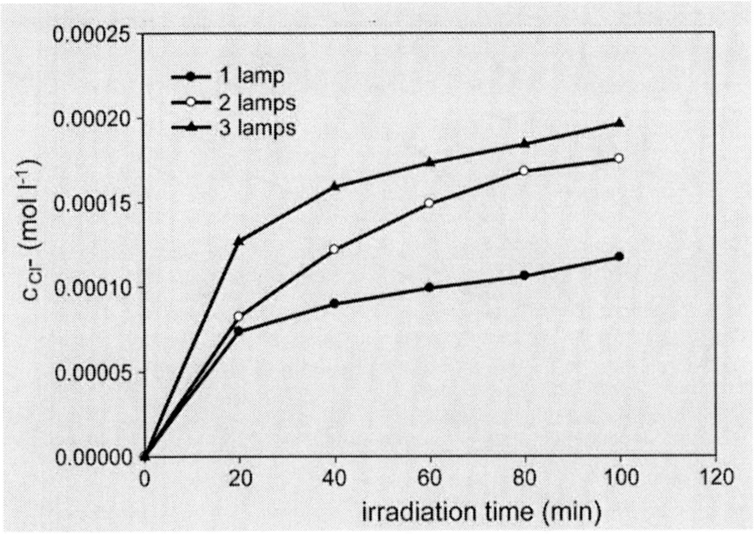

Figure 20. Effect of number of titania-coated EDLs on MCAA photodegradation (EDL: two layers; 90 °C).

6.1.4. Effect of Pure and V-, Zr-, and Ag-doped Titania (in Batch Type)

The photocatalytic activity (RhB decomposition, see chapter 3.4.) of doped TiO_2 was greatly enhanced by incorporating it with V, Zr, and Ag. Therefore, Hg-EDLs were then coated with a thin film of these doped titania samples and used for the photocatalysis of MCAA in a MW field. Figure 21 shows the increase in chloride ion concentration in an aqueous solution in the presence of V, Zr, and Ag-doped TiO_2 under light irradiation (EDL lines: 365, 405, and 436 nm). Obviously, the photocatalytic activity of Ag(3%)/TiO_2 (Figure 21, open triangle) was superior for the decomposition of MCAA in a MW field than that of the undoped counterpart.

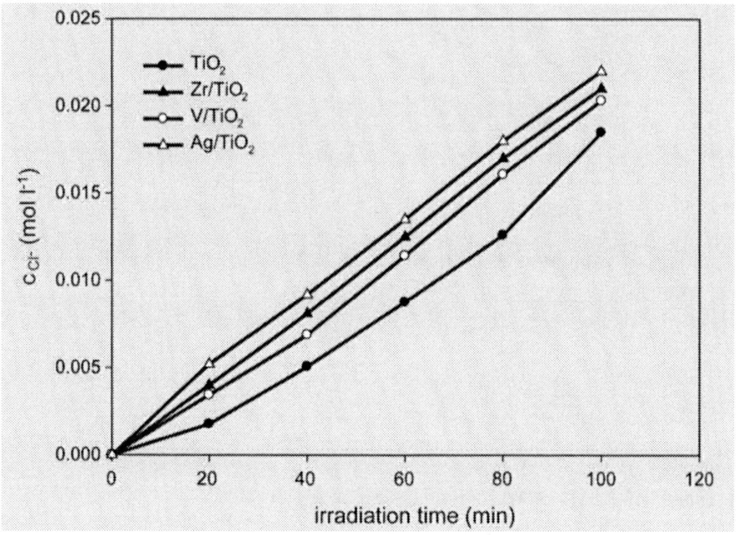

Figure 21. Effect of pure and Zr, V, and Ag-doped TiO_2 on the photocatalytic decomposition of MCAA (EDL intensity: 5.56 μW cm^{-2}; EDL: two layers).

We also investigated the effect of doping amount of vanadyl acetylacetonate on the photocatalytic activity of V(3%)/TiO_2 (1.167 at.%). When the doping level surpasses the optimum, which usually lies at a very low dopant concentration and low visible light absorbance, dopants become recombination sites and have adverse effects on photocatalysis [168]. We have found that the other two doped TiO_2 catalysts (1 and 5%) with different amounts of vanadium possessed similar morphologies (0.459 and 1.778 at.%) and showed different photocatalytic activities for the MCAA degradation (Figure 22). The vanadium percentage (at.%)

was evaluated by the XPS. The best degradation ability was obtained in the $V(5\%)/TiO_2$ photocatalyst (Figure 22, solid triangle).

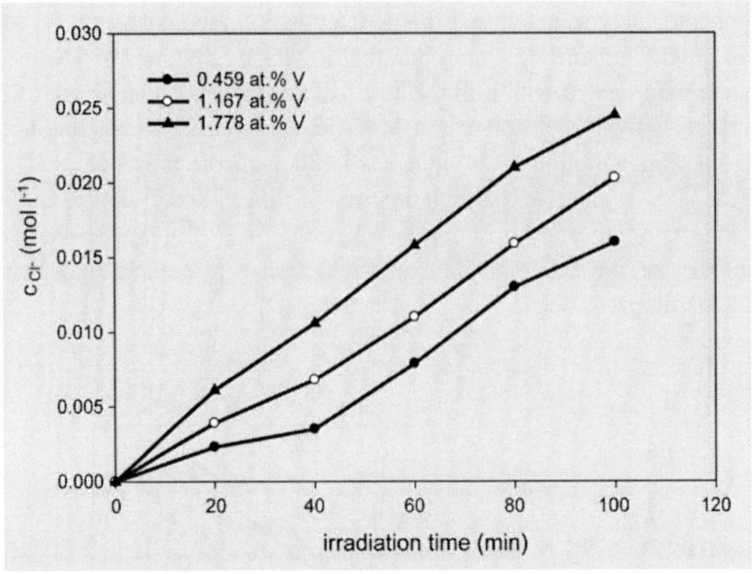

Figure 22. Effect of vanadium amount on the MCAA photocatalytic efficiency of doped TiO_2 (EDL intensity: 5.56 μW cm^{-2}; EDL: two layers).

6.2. Effect of Reaction Conditions

6.2.1. Effect of Initial pH (in Batch Type)

The pH has a complex effect on the rates of photocatalytic reaction and the observed effect generally depends on the type of the pollutant and semiconductor. It is also known [144] that pH has an effect on the electrostatic charge of the TiO_2 surface, which determines the density of $TiOH_2^+$ groups.

In experiment, the pH was adjusted before photocatalytic reaction by addition of the corresponding amount of NaOH to the solution. As presented in Figure 23, the reaction was enhanced in an alkaline medium. The reason for the increase to the high initial pH is having more hydroxide ions present that may produce more hydroxyl radicals ($h^+_{vb} + OH^- \rightarrow OH^\bullet$). The presence of anions, e. g. chloride anion, can affect the degradation process.

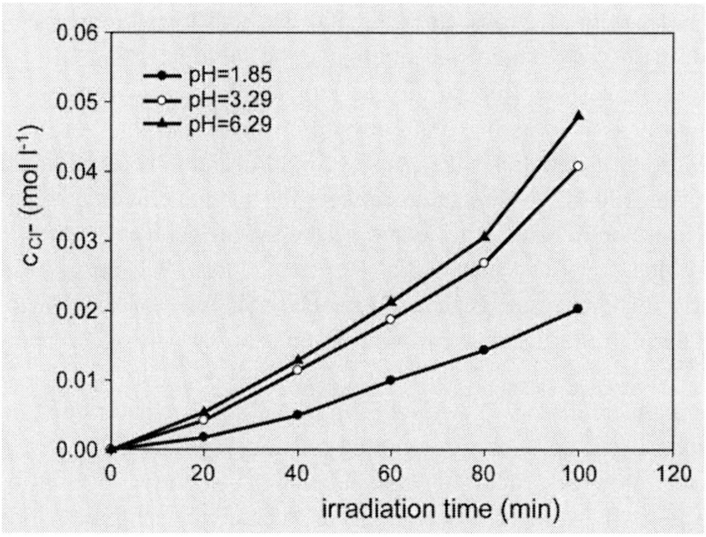

Figure 23. Effect of initial pH on the photocatalytic decomposition of MCAA (EDL intensity: 5.56 μW.cm^{-2}; EDL: 2 layers).

These anions can react with hydroxyl radicals and consume them ($OH^{\bullet} + Cl^- \rightarrow Cl^{\bullet} + OH^-$) and they can absorb UV radiation as well.

The pH value may affect the surface charge of the catalyst. At lower initial pH the degradation rate decreases in the presence of chloride ions. It is because of the adsorption on the positively charged catalyst surface. However, at high initial pH, there is no significant influence of chloride ions [77].

6.2.2. Effect of Air Bubbling (in Batch and Continuous-flow Types)

Air bubbling is the simplest way to supply the appropriate amount of oxygen to the reaction mixture. The presence of oxygen as the electron acceptor is recommended so as to prevent the recombination between the generated positive holes and electrons. The dioxygen seems to participate in the degradation process from the initial stages of the MCAA photocatalysis by reacting rapidly with the intermediate radicals formed [161].

Horikoshi [126,144] found that under a low oxygen concentration (nitrogen-purging) the photodegradation rate in the presence of microwaves was higher than that in the absence of MW irradiation. As mentioned above (chapter 4), specific interactions of microwaves with the UV-illuminated TiO_2 particle surface give rise to the generation of additional surface defects that can prohibit the recombination of holes with electrons. In EDL/TiO_2 the high transference rate decreases the need of dissolved oxygen concentration in the solution.

In experiments the effect of air gas bubbling was also checked. As can be seen in Figure 24, the presence of air doesn't influence the photodecomposition of MCAA. More to the contrary, we can see a mild decrease because this reaction is carried out at 100 °C. Temperature affects the amount of oxygen that can be dissolved in water. When fresh water is heated up, air bubbles start to form. If the water is cooled down and then again reheated, bubbles will not appear until the water starts to boil (water is deaerated). The solubility of oxygen in water at 101.325 kPa pressure decreases from 8.06 ppm at 25 °C to 0.87 ppm at 95 °C [169].

The amount of dissolved oxygen is an important parameter due to its limited concentration in the solution. As reported previously (Figure 24), in a batch type photoreactor the presence of air does not significantly influence the decomposition of MCAA. Therefore, a photocatalytic continuous-flow reactor with re-circulation was used to investigate the air bubbling effect on MW-assisted MCAA photocatalysis. The reaction mixture was bubbled in the glass reservoir at the laboratory temperature and then circulated through the photoreactor at 90 °C. The differences in performance between air bubbled through the mixture and no air bubbled through the mixture are presented in Figure 25. These results indicated the important role of oxygen in the photocatalytic reaction of MCAA, as was presented in previous papers [158,161,162,170].

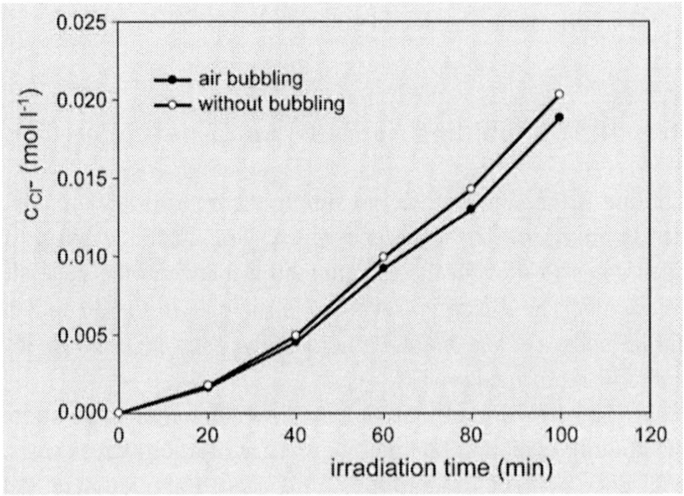

Figure 24. Time dependence of c_{Cl^-} on an air bubbling of the reaction mixture (EDL intensity: 5.56 μW.cm^{-2}; EDL: 2 layers).

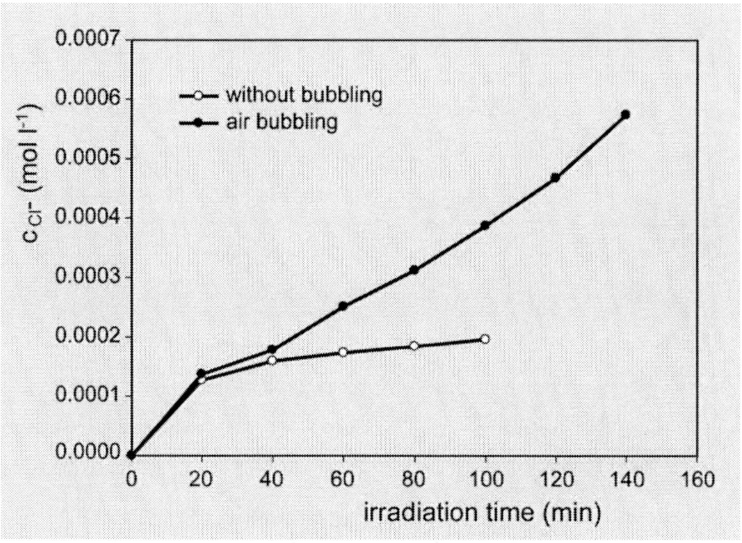

Figure 25. Time dependence of c_{Cl^-} with and without air bubbling through the reaction mixture (three EDLs: two layers; reaction temperature 90 °C).

6.2.3. Effect of H_2O_2 Dosages (in Batch Type)

One effective strategy for inhibiting e^-/h^+ recombination is to add another electron acceptor to the reaction system. The use of H_2O_2 [155] has been demonstrated to enhance the rates of Brilliant Red X-3B degradation remarkably because it traps the photogenerated electrons more efficiently than molecular oxygen. Secondly, the application of H_2O_2 produces •OH radicals (at the conduction band [143]) which can again degrade some pollutants.

The photocatalytic decomposition of MCAA was increased, when H_2O_2 was present in the solution. The effect of H_2O_2 was enhanced by increasing the H_2O_2 dose from 0 to 0.3% v/v. At the initial stage, the reaction rate was increased abruptly, then went to a balance after the H_2O_2 dosage was 0.1% v/v (Cl^- concentration 0.021175 mol.l^{-1}; reaction time: 100 min; the EDL intensity: 5.60 μW.cm^{-2}; EDL: 2 layers). The effect of optimal H_2O_2 dosage 0.1% v/v on photocatalytic degradation of MCAA was shown in Figure 26. It has been observed that H_2O_2 dosage has only a slight effect on the reaction enhancement.

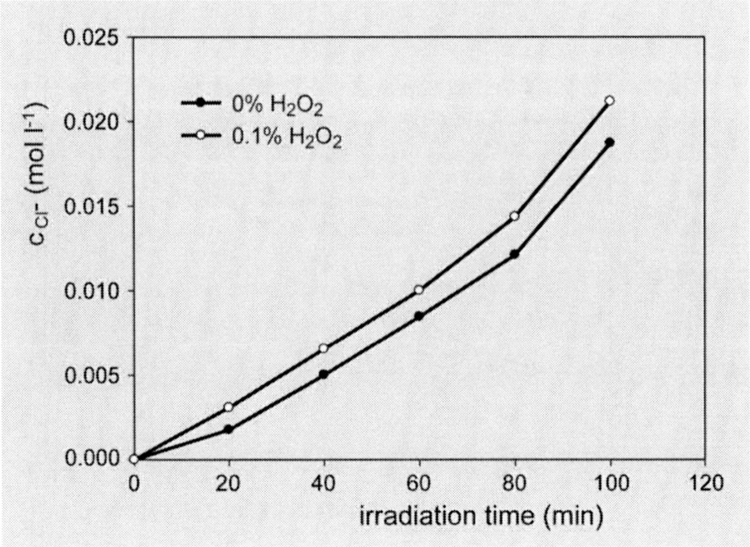

Figure 26. Effect of adding H_2O_2 on photocatalytic decomposition of MCAA (EDL intensity: 5.60 μW.cm^{-2}; EDL: 2 layers).

6.2.4. Effect of Flow Rate and Reaction Temperature (in Continuous-flow Type)

It was not easy to carry out MW assisted photoreactions at different temperatures. The reaction temperature was mostly assigned as the boiling point of the solvent and some superheating may have occurred. The influence of superheating on reaction behavior is well known. Superheating can increase reaction temperature and consequently, the reaction rate.

The effect of volumetric flow rate on the reaction temperature ($T = f(V)$) is summarized in Figure 27. The results indicate that increasing the flow rate from 5 to 15 l h^{-1} led to a rapid decrease in reaction temperature (from 95 °C to 40 °C), however, at higher flow rate from 15 to 70 l h^{-1}, the reaction temperature decreased minimally (from 40 °C to 35 °C) and was almost independent on the volumetric flow rate. These results are logical, because with increasing flow rate (15-70 l h^{-1}) the residence time of the reactant (aqueous MCAA) decreases in the microwave photocatalytic reactor and the reaction mixture is warmed up by hot EDLs. In the case of low flow rate (5 l h^{-1}) the highly MW absorbing water was heated mainly by MW radiation.

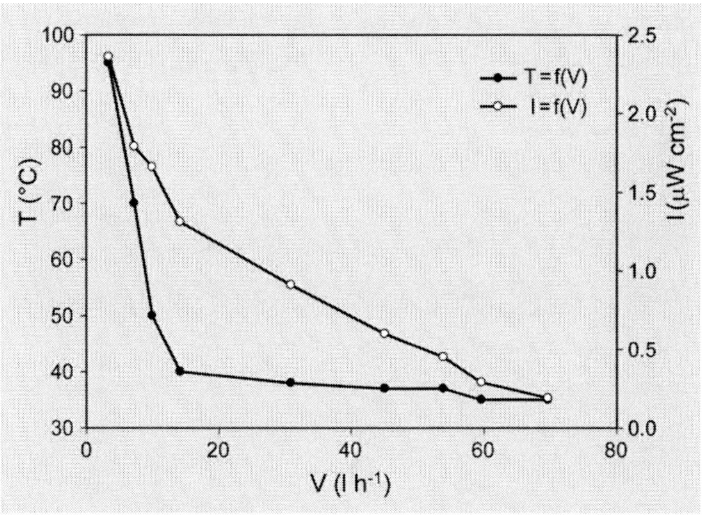

Figure 27. Effect of volumetric flow rate on reaction temperature ($T = f(V)$) and on the light intensity of EDLs (366 nm line) ($I = f(V)$) (EDL: two layers).

The effect of volumetric flow rate on the light intensity of EDLs (for the 366 nm line) ($I = f(V)$) is also plotted in Figure 27. The results indicate that with increasing flow rate from 5 to 70 l h^{-1}, the light intensity of EDLs decreases continuously (from 2.4 μW cm^{-2} to 0.2 μW cm^{-2}). Therefore too high a volumetric flow rate is inconvenient, giving low EDL light intensity and low reaction temperature (*vide infra*).

The temperature of the MCAA aqueous medium affects the EDL envelope (Pyrex) temperature and therefore the plasma lighting properties [13]. The relative intensities of the emission peaks in Hg-EDL were found to be very dependent on temperature [26]; the 254 nm line was suppressed with increasing temperature, however, the 366 nm line (titania-coated Hg-EDL on Pyrex) was enhanced. This phenomenon was disclosed in our paper [26] and Figure 28 shows the effect of reaction temperature on EDLs light intensity for one and three EDLs. The results show that the highest intensity values I (μW cm^{-2}) were obtained at higher temperatures.

The investigation involving the influence of temperature on photocatalytic MCAA reaction was made in the range since 35 to 90 °C when the reaction temperature was altered by changing feed flow rate. Figure 29 shows the dependence of Cl^- concentration on the reaction temperature over time. The Cl^- concentration was maximum at 90 °C indicating that intensive heating is required for degradation of the MCAA in the photo catalytic reaction. Therefore, the photoreactor flow rate significantly affects the apparent photodegradation rate of MCAA. A very small amount of chloride ions (0.00012 mol l^{-1}) were generated to ensure an ultra-short contact time of reaction mixture with one EDL. For practical purposes it would be necessary to use more EDLs.

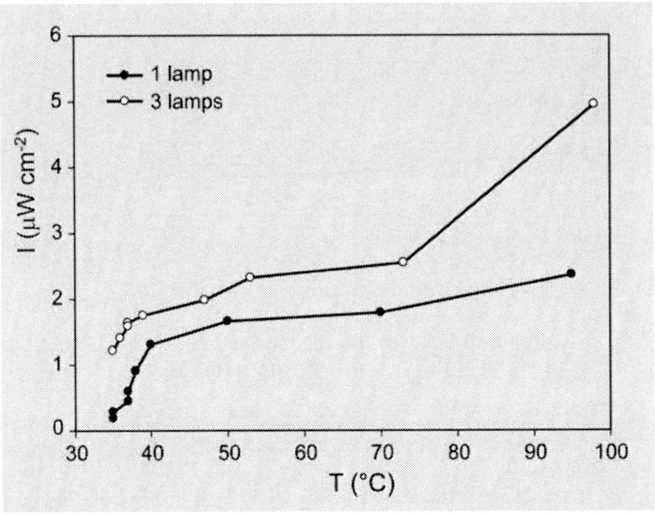

Figure 28. Effect of reaction temperature on the light intensity of EDLs (EDL: two layers).

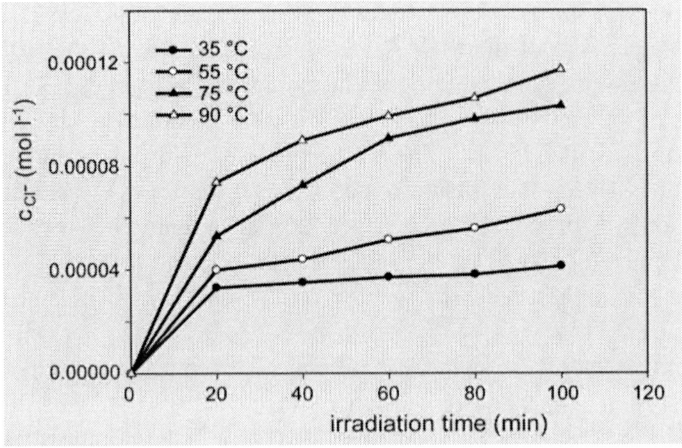

Figure 29. Dependence of chloride ion concentration on the reaction temperature over time (EDL: two layers).

Chapter 7

CONCLUSION

In this review, we present a new method for carrying out photocatalytic reactions with high efficiency in the decomposition of mono-chloroacetic acid (MCAA). This special film arrangement of the photocatalyst form on the electrodeless discharge lamp (EDL) and the microwave source of UV/Vis light provide a new form of the TiO_2 utilization.

Theory of the microwave discharge in EDLs, their construction, spectral characteristics and performance were described. The preparation (sol-gel process and dip-coating) and characterization (XRD, Raman, XPS, SEM, AFM, UV/Vis) of pure titania and titania-doped (Zr, V, Cr, Mn, Fe, Co, Ni, and Ag) thin films on EDLs was mentioned in details. The photocatalytic activity of the prepared thin titania films was examined by the decomposition test of Rhodamine B. The interaction of microwave radiation with the UV/Vis-illuminated titania was briefly commented. Novel microwave photocatalytic reactors in the batch and the continuous-flow experimental set-up were described. The *microwave photocatalysis* of MCAA with titania-coated EDLs was performed in details. MCAA was totally decomposed to HCl, CO_2, and H_2O over the irradiated titania film in microwave field.

We reported the effect of EDLs (number of coating cycles, light intensity, number of coated lamps, and titania-doped film) and reaction conditions (initial pH value, air bubbling, H_2O_2 dosage, flow rate, and reaction temperature) on the photocatalytic degradation of MCAA on thin titania-coated EDLs in a batch microwave photocatalytic reactor and in a continuous-flow microwave reactor.

The batch experimental set-up study revealed that the photocatalytic reaction efficiency depends on the intensity of light and initial pH value of the solution. Moreover, the degradation of MCAA was enhanced in an alkaline solution and in the presence of H_2O_2, and significantly enhanced by increasing the intensity of light. Furthermore, this review also discloses that efficiency of UV/Vis irradiation is not influenced by the number of coating cycles or by air bubbling. MCAA also yielded no observed organic intermediates under these thermal reaction conditions.

The continuous-flow experimental set-up study revealed that the photocatalytic reaction temperature and light intensity of the EDLs depend inversely on the flow rate, but that the 366 nm line intensity of the EDL is directly proportional to the reaction temperature. Moreover, the photodegradation of MCAA was enhanced by heating and significantly enhanced by bubbling air through the reaction mixture in the glass reservoir at laboratory temperature. Furthermore, the photocatalytic efficiency was increased with the number of the titania-coated EDLs inserted in the glass tube reactor.

It has been found that with a suitable amount of V, Zr, and Ag showing significant absorption in the visible region, the metal dopants effectively increase the photocatalytic activity of the TiO_2 film on the EDLs. The best apparent degradation rate constant (0.0125 min^{-1}), which was higher than that on the pure TiO_2 film by 1.7 times, was obtained on the Ag(3%)/TiO_2 photocatalyst. The doping amount effect of vanadium acetylacetonate on the photocatalytic efficiency of V-doped TiO_2 has been determined.

In this review, we have discussed how the concept of microwave-assisted photocatalysis on the titania thin films has already became an important issue in chemistry. Although still in the beginning, detailed analysis of past and present literature confirms explicitly the usefulness of this method of chemical activation. The application of the titania coated EDL simplifies the technical procedure, especially in the field of organic photochemical and photocatalytic synthesis, environmental chemistry, or analysis.

ACKNOWLEDGMENT

Thanks are due to the Grant Agency of the Czech Republic (104/07/1212) for funding this research. In addition, H.Ž. thanks the Grant Agency of the Czech Republic for part funding her PhD studentship (Doctoral Grant No. 203/08/H032). We are also grateful to Milestone s.r.l. (Italy) for their technical support.

REFERENCES

[1] Herrmann, J.-M. *Top. Catal.* 2005, *34*, 49-65.
[2] Zhao, J.; Yang, X. *Build. Environ.* 2003, *38*, 645-654.
[3] Rajeshwar, K.; Osugi, M. E.; Chanmanee, W.; Chenthamarakshan, C. R.; Zanoni, M. V. B.; Kajitvichyanukul, P.; Krishnan-Ayer, R. *J. Photochem. Photobiol. C: Photochem. Rev.* 2008, *9*, 171-192.
[4] Gedye, R.; Smith, F.; Westaway, K.; Ali, H.; Baldisera, L.; Laberge, L.; Rousell, J. *Tetrahedron Lett.* 1986, *27*, 279-282.
[5] de la Hoz, A; Diaz-Ortiz, A.; Moreno, A. *Chem. Soc. Rev.* 2005, *34*, 164-178.
[6] Lidström, P.; Tierney, J.; Wathey, B.; Westman, J. *Tetrahedron* 2001, *57*, 9225-9283.
[7] Perreux L.; Loupy, A. *Tetrahedron* 2001, *57*, 9199-9223.
[8] Maeda, M.; Amemiya, H. *New J. Chem.* 1995, *19*, 1023-1028.
[9] Chemat, F.; Poux, M.; DiMartino, J. L.; Berlan, J. *J. Microw. Power Electromagn. Energy* 1996, *31*, 19-22.
[10] Cravotto G., Beggiato, M.; Penoni, A.; Palmisano, G.; Tollari, S.; Leveque, J. M.; Bonrath, W. *Tetrahedron Lett.* 2005, *46*, 2267-2271.
[11] Cutress, I. J.; Marken, F.; Compton, R. G. *Electroanalysis* 2009, *21*, 113-123.
[12] Církva, V.; Hájek, M. *J. Photochem. Photobiol. A: Chem.* 1999, *123*, 21-23.
[13] Klán, P.; Církva, V. In *Microwaves in Organic Synthesis*, Loupy, A.; Ed.; Microwaves in Photochemistry, Wiley-WCH: Weinheim, GE, 2006; Vol. 2, pp 860-897.
[14] Horikoshi S.; Hidaka, H.; Serpone, N. *Environ. Sci. Technol.* 2002, *36*, 1357-1366.

[15] Horikoshi, S.; Tokunaga, A.; Watanabe, N.; Hidaka, H.; Serpone, N. *J. Photochem. Photobiol. A: Chem.* 2006, *177*, 129-143.
[16] Fehsenfeld, F. C.; Evenson, K. M.; Broida, H. P. *Rev. Sci. Instr.* 1965, *36*, 294-298.
[17] Brown, S. C. *Introduction to Electrical Discharges in Gases*, John Wiley and Sons: NY, US, 1966.
[18] Wilkinson, P. G.; Tanaka, Y. *J. Opt. Soc. Am.* 1955, *45*, 344-349.
[19] Chandler, R.; Popov, O.; Shapiro, E. K.; Maya, J. *U.S. Pat. Appl.* (2001) US 2001000941.
[20] Ganeev, A.; Gavare, Z.; Khutorshikov, V. I.; Khutorshikov, S. V.; Revalde, G.; Skudra, A.; Smirnova, G. M.; Stankov, N. R. *Spectrochim. Acta B* 2003, *58*, 879-889.
[21] Wertheimer, M. R.; Fozza, A. C.; Hollander, A. *Nucl. Instrum. Methods Phys. Res. Sect. B-Beam Interact. Mater. Atoms* 1999, *151*, 65-75.
[22] Gleason, W. S.; Pertel, R. *Rev. Sci. Instr.* 1971, *42*, 1638-1643.
[23] Haarsma, J. P. S.; De Jong, G. J.; Agterdenbos, J. *Spectrochim. Acta B* 1974, *29*, 1-18.
[24] Sneddon, J.; Browner, R. F.; Keliher, P. N.; Winefordner, J. D.; Butcher, D. J.; Michel, R. G. *Prog. Analyt. Spectrosc.* 1989, *12*, 369-402.
[25] Církva, V.; Vlková, L.; Relich, S.; Hájek, M. *J. Photochem. Photobiol. A: Chem.* 2006, *179*, 229-233.
[26] Müller, P.; Klán, P.; Církva, V. *J. Photochem. Photobiol. A: Chem.* 2003, *158*, 1-5.
[27] Müller, P.; Klán, P.; Církva, V. *J. Photochem. Photobiol. A: Chem.* 2005, *171*, 51-57.
[28] Phillips, R. *Sources and Applications of Ultraviolet Radiation*; Academic Press: London, UK, 1983.
[29] Spero, D. M.; Matthews, J. C. *J. Radiat. Curing* 1979, *6*, 6-10.
[30] Dakin, J. T.; Berry, T.; Duffy, M. E.; Russell, T. D. *U.S. Pat. Appl.* (1994) US 5363015.
[31] Dolan, J. T.; Ury, M. G.; Turner, B. P.; Waymouth, J. F.; Wood, C. H. *PCT Int. Appl.* (1993) WO 9321655.
[32] Dolan, J. T.; Turner, B. P.; Ury, M. G.; Wood, C. H. *U.S. Pat. Appl.* (1997) US 5682080.
[33] Dolan, J. T.; Ury, M. G.; Wood, C. H. *US Pat. Appl.* (1995) US 5404076.
[34] Johnson, P. D.; Dakin, J. T.; Anderson, J. M. *U.S. Pat. Appl.* (1989) US 4810938.
[35] Shea, A. J.; Feuersanger, A. E.; Keeffe, W. M.; Struck, C. W. *Eur. Pat. Appl.* (1994) EP 0603014.

[36] Russell, T. D.; Berry, T.; Dakin, J. T.; Duffy, M. E. *Eur. Pat. Appl.* (1993) EP 0542467.

[37] Marshall, G. B.; West, T. S. *Anal. Chim. Acta* 1970, *51*, 179-190.

[38] Patel, B. M.; Browner, R. F.; Winefordner, J. D. *Anal. Chem.* 1972, *44*, 2272-2277.

[39] Browner, R. F.; Winefordner, J. D.; Glenn, T. H. *U.S. Pat. Appl.* (1974) US 3786308.

[40] Ury, M. G.; Wood, C. H. *U.S. Pat. Appl.* (1989) US 4859906.

[41] Florian, D.; Knapp, G. *Anal. Chem.* 2001, *73*, 1515-1520.

[42] Ocana, M.; Dodnes, V.; García Ramos, J. V.; Serna, C. J. J. *Solid State Chem.* 1998, *75*, 364-372.

[43] Ocana, M.; Serna, C. J. *Spectrochimica Acta* 1991, *47A*, 765-774.

[44] Byen, D.; Jin, Y.; Kim, B. *J. Hazard. Mater.* 2000, *73*, 199-206.

[45] Waters, J. P.; Smyth-Boyle, D.; Govender, K.; Green, A.; Durrant, J.; O'Brien, P. *Chem. Vap. Deposition* 2005, *11*, 254-260.

[46] Martinu, L.; Poitras, D. *J. Vac. Sci. Technol. A* 2000, *18*, 2619-2645.

[47] Frach, P.; Glöb, D.; Metzner, C.; Modes, T.; Scheffel, B.; Zywitzki, O. *Vacuum* 2006, *80*, 679-683.

[48] Scolan, A.; Sánchez, C. *Chem. Mater.* 1998, *10*, 3217-3223.

[49] Pedraza, F.; Vazquez, A. *J. Phys. Chem. Solids* 1999, *60*, 445-448.

[50] Kim, C.-S.; Moon, B. K.; Park, J.-H.; Chung, S. T; Son, S.-M. *J. Cryst. Growth* 2003, *254*, 405-410.

[51] Lee, S.-H.; Kang, M.; Cho, S. M.; Han, G. Y.; Kim, B.-W.; Yoon, K. J.; Chung, C.-H. *J. Photochem. Photobiol. A: Chem.* 2001, *146*, 121-128.

[52] Juan, Z.; Li, B.; Zhang, J.; Xu, C. *J. Sol-Gel Sci. Techn.* 2006, *39*, 249-253.

[53] Yu, J.; Zhao, X.; Zhao, Q. *Mater. Chem. Phys.* 2001, *69*, 25-29.

[54] Reddy, K. M.; Reddy, C. V. G.; Panorama, S. V. *J. Solid State Chem* 2001, *158*, 180-186.

[55] Gelover, S.; Mondragón, P.; Jiménez, A. *J. Photochem. Photobiol. A: Chem.* 2004, *165*, 241-246.

[56] Su, C.; Hong, B.-Y.; Tseng, C.-M. *Catal. Today* 2004, *96*, 119-126.

[57] Sopyan, I.; Watanabe, M.; Murasawa, S.; Hashimoto, K.; Fujishima, A. *J. Photochem. Photobiol. A: Chem.* 1996, *98*, 79-86.

[58] Hung, C.-H.; Marinas, B. I. *Environ. Sci. Technol.* 1997, *31*, 1440-1445.

[59] Noguchi, T.; Fujishima, A. *Environ. Sci. Technol.* 1998, *32*, 3831-3833.

[60] Kluson, P.; Luskova, H.; Cajthaml, T.; Solcova, O. *Thin Solid Films* 2006, *495*, 18-23.

[61] Kluson, P.; Kacer, P.; Cajthaml, T.; Kalaji, M. *J. Mater. Chem.* 2001, *11*, 644-651.

[62] Kluson, P.; Kacer, P.; Cajthaml, T.; Kalaji, M. *Chem. Biochem. Eng. Q.* 2003, *17*, 183-190.

[63] Hamal, D. B.; Klabunde, K. J. *J. Colloid Interface Sci.* 2007, *311*, 514-522.

[64] Kim, S.; Hwang, S.-J.; Choi, W. *J. Phys. Chem.* 2005, *109*, 24260-24267.

[65] Chen, J.; Yao, M.; Wang, X. *J. Nanopart. Res.* 2008, *10*, 163-171.

[66] Di Paola, A.; Marci, G.; Palmisano, L.; Schiavello, M.; Upsali, K.; Ikeda, S.; Ohtani, B. *J. Phys. Chem. B* 2002, *106*, 637-645.

[67] Chen, D. M.; Yang, D.; Geng, J. Q. *Appl. Surf. Sci.* 2008, *255*, 2879-2884.

[68] Lin, C. F.; Wu, C. H.; Onn, Z. N. *J. Hazard. Mater.* 2008, *154*, 1033-1039.

[69] Do, Y. R.; Lee, W.; Dwight, K.; Wold, A. *J. Solid State Chem.* 1994, *108*, 198-201.

[70] Liu, Z. L.; Cui, Z. L.; Zhang, Z. K. *Mater. Charact.* 2005, *54*, 123-129.

[71] Pérez-Hernandéz, R.; Mendoza-Anaya, D.; Fernandéz, M. E.; Gómez-Cortés, A. *J. Mol. Catal. A* 2008, *281*, 200-206.

[72] Serpone, N.; Lawless, D. *Langmuir* 1994, *10*, 643-652.

[73] Sharma, S. D.; Saini, K. K.; Kant, C.; Sharma, C. P.; Jain, S. C. *Appl. Catal. B* 2008, *84*, 233-240.

[74] Zou, X. X.; Li, G. D.; Guo, M. Y.; Li, X. H.; Liu, D. P.; Su, J.; Chen, J. S. *Chem. Eur. J.* 2008, *14*, 11123-11131.

[75] Asahi, R.; Morikawa, T.; Ohwaki, T.; Aoki, K.; Taga, Y. *Science* 2001, *293*, 269-271.

[76] Moon, J.; Takagi, H.; Fujishiro, Y.; Awano, M. *J. Mater. Sci.* 2001, *36*, 949-955.

[77] Kaneko, M.; Okura, I. *Photocatalysis. Science and Technology*; Springer: Tokyo, JP, 2002.

[78] Di Paola, A.; García-Lopéz, E.; Ikeda, S.; Marcí, G.; Ohtani, B.; Palmisano, L. *Catal. Today* 2002, *75*, 87-93.

[79] Masahashi, N.; Oku, M. *Appl. Surf. Sci.* 2008, *254*, 7056-7060.

[80] Ambrus, Z.; Balázs, N.; Alapi, T.; Wittmann, G.; Sipos, P.; Dombi, A.; Mogyorósi, K. *Appl. Catal. B* 2008, *81*, 27-37.

[81] Alexandrescu, R.; Morjan, I.; Scarisoreanu, M.; Birjega, R.; Popovici, E.; Soare, I.; Gavrila-Florescu, L.; Vlivu, I.; Sandu, I.; Dumitrache, F.; Prodan, G.; Vasile, E.; Figgemeier, E. *Thin Solid Films* 2007, *515*, 8438-8445.

[82] Subramanian, M.; Vijayalakshmi, S.; Venkataraj, S.; Jayavel, R. *Thin Solid Films* 2008, *516*, 3776-3782.

[83] Maensiri, S.; Laokul, P.; Klinkaewnarong, J. *J. Magn. Magn. Mater.* 2006, *302*, 448-453.

[84] Martínez, T. L. M.; Montes de Correa, C.; Odriozola, J. A.; Centeno, M. A. *J. Mol. Catal. A* 2006, *253*, 252-260.

[85] Mohamed, M. M.; Otoman, I.; Mohamed, R. M. *J. Photochem. Photobiol. A: Chem.* 2007, *191*, 153-161.

[86] Zhang, S.; Chen, Y.; Yu, Y.; Wu, H.; Wang, S.; Zhu, B.; Juany, W.; Wu, S. *J. Nanopart. Res.* 2008, *10*, 871-875.

[87] Wu, C.-G.; Chao, C.-C.; Kuo, F.-T. *Catal. Today* 2004, *97*, 103-112.

[88] Wu, J. C. S.; Chen, C. H. *J. Photochem. Photobiol. A: Chem.* 2004, *163*, 509-515.

[89] Tsuyumoto, I.; Nawa, K. *J. Mater. Sci.* 2008, *43*, 985-988.

[90] Li, Z. H.; Dong, H.; Zhang, Y. F. *J. Phys. Chem. C* 2008, *112*, 16046-16051.

[91] Du, Y. K.; Gan, Y. Q.; Yang, P.; Zhao, F.; Hua, N. P.; Juany, L. *Thin Solid Films* 2005, *491*, 133-136.

[92] Garzella, C.; Comini, E.; Bontempi, E.; Depero, L. E.; Frigari, C.; Sterveglieri, G. *Sens. Actuators B* 2002, *83*, 230-237.

[93] Wang, Y.; Hao, Y.; Cheby, H.; Ma, J.; Xu, B.; Li, W.; Cai, S. *J. Mater. Sci.* 1999, *34*, 2773-2779.

[94] Venkatachalam, N.; Palanichamy, M.; Arabindoo, B.; Murugesan, V. *J. Mol. Catal. A* 2007, *266*, 158-165.

[95] Lippens, P. E.; Chadwick, A. V.; Weibel, A.; Bouchet, R.; Knauth P. *J. Phys. Chem. C* 2008, *112*, 43-47.

[96] Arabatzis, I. M.; Stergiopoulos, T.; Bernard, M. C.; Labou, D.; Neophytides, S. G.; Falaras, P. *Appl. Catal. B* 2003, *42*, 187-201.

[97] Hamal, D. B.; Klabunde, K. J. *J. Colloid Interface Sci.* 2007, *311*, 514-522.

[98] Epifani, M.; Giannini, C.; Tapfer, L.; Vasanelli, L. *J. Am. Ceram. Soc.* 2000, *10*, 2385-2393.

[99] He, C.; Yu, Y.; Hu, X.; Larbot, A. *Appl. Surf. Sci.* 2002, *200*, 239-247.

[100] Martin, S. T.; Kortison, C. L.; Hoffmann, M. R. *J. Phys. Chem.* 1994, *98*, 13695-13704.

[101] Ambrus, Z.; Balázs, N.; Alapi, T.; Wittmann, G.; Sipos, P.; Dombi, A.; Mogyorósi, K. *Appl. Catal. B* 2008, *81*, 27-37.

[102] Ohsaka, T.; Izumi, F.; Fujiki, Y. *J. Raman Spectrosc.* 1978, *7*, 321-324.

[103] Tompsett, G. A.; Bowmaker, G. A.; Cooney, R. P.; Metson, J. B.; Rodgers, K. A.; Seakins, J. M. *J. Raman Spectrosc.* 1995, *26*, 57-62.

[104] Joung, S. K.; Amemiya, T.; Murabayashi, M.; Itoh, K. *Chem. Eur. J.* 2006, *12*, 5526-5534.

[105] Benjaram, M. R.; Chowdhury, B.; Reddy, E. P.; Fernandez, A. *J. Mol. Catal. A* 2000, *162*, 431-441.

[106] Yu, G. H.; Chai, C. L.; Zhu, F. W.; Xiao, J. M.; Lai, W. Y. *Appl. Phys. Lett.* 2001, *78*, 1706-1708.

[107] Hung, W. C.; Chen, Y. C.; Chu, H.; Tseng, T. K. *Appl. Surf. Sci.* 2008, *255*, 2205-2213.
[108] [Amadelli, R.; Samiolo, L.; Maldotti, A.; Molinari, A.; Valigi, M.; Gazzoli, D. *Int. J. Photoenergy* 2008, article ID 853753, 9 pp.
[109] Sharma, S. D.; Saini, K. K.; Kant, C.; Sharma, C. P.; Jain, S. C. *Appl. Catal. B* 2008, *84*, 233-240.
[110] Yang, X.; Cao, C.; Hohn, K.; Ericsson, L.; Maghirang, R.; Hamal, D.; Klabunde, K. *J. Catal.* 2008, *252*, 296-302.
[111] Venezia, A. M.; Palmisano, L.; Schiavello, M.; Martin, C.; Martin, I.; Rives, V. *J. Catal.* 1994, *147*, 115-122.
[112] Davydov, L.; Reddy, E. P.; France, P.; Smirniotis, P.G. *J. Catal.* 2001, *203*, 157-167.
[113] Ao, Y.; Xu, J.; Fu, D.; Yuan, C. *J. Phys. Chem. Solids* 2008, *69*, 2660-2664.
[114] Chen, D. A.; Bartelt, M. C.; Seutter, S. M.; McCarty, K. F. *Surf. Sci.* 2000, *464*, L708-L714.
[115] Vitos, L.; Ruban, A. V.; Skriver, H. L.; Kollár, J. *Surf. Sci.* 1998, *411*, 186-202.
[116] Nagaveni, K.; Hegde, M. S.; Ravishankar, N.; Subbanna, G. N.; Madras, G. *Langmuir* 2004, *20*, 2900-2907.
[117] Chen, C. Y.; Tuan, W. H. *J. Am. Ceram. Soc.* 2000, *83*, 2988-2992.
[118] Qourzal, S.; Tamimi, M.; Assabbane, A.; Bouamrane, A.; Nounah, A.; Laânab, L.; Ait-Ichou, Y. *J. Appl. Sci.* 2006, *6*, 1553-1559.
[119] Ao, Y.; Xu, J.; Fu, D.; Yuan, C. *Appl. Surf. Sci.* 2008, *255*, 3137-3140.
[120] Rashidzadeh, M. *Int. J. Photoenergy* 2008, article ID 245981, 4 pp.
[121] Chao, H. E.; Yun, Y. U.; Xingfang, H. U.; Larbot, A. *J. Eur. Ceram. Soc.* 2003, *23*, 1457-1464.
[122] Horikoshi, S.; Hidaka, H.; Serpone, N. *J. Photochem. Photobiol. A: Chem.* 2004, *161*, 221-225.
[123] Horikoshi, S.; Kajitani, M.; Sato, S.; Serpone, N. *J. Photochem. Photobiol. A: Chem.* 2007, *189*, 355-363.
[124] Sur, U. K.; Marken, F.; Compton, R. G.; Coles, B. A. *New J. Chem.* 2004, *28*, 1544-1549.
[125] Booske, J. H.; Cooper, R. F.; Dobson, I. *J. Mater. Res.* 1992, *7*, 495-501.
[126] Horikoshi, S.; Hidaka, H.; Serpone, N. *J. Photochem. Photobiol. A: Chem.* 2003, *159*, 289-300.
[127] Emeline, A. V.; Kuzmin, G. N.; Purevdorj, D.; Ryabchuk, V. K.; Serpone, N. *J. Phys. Chem. B* 2000, *104*, 2989-2999.
[128] Horikoshi, S.; Hidaka, H.; Serpone, N. *Chem. Phys. Lett.* 2003, *376*, 475-480.

[129] Blount, M. C.; Kim, D. H.; Falconer, J. L. *Environ. Sci. Technol.* 2001, *35*, 2988-2994.
[130] Martyanov, I. N.; Klabunde, K. J. *J. Catal.* 2004, *225*, 408-416.
[131] Klán, P.; Literák, J. *Collect. Czech. Chem. Commun.* 1999, *64*, 2007-2018.
[132] Klán, P.; Hájek, M.; Církva, V. *J. Photochem. Photobiol. A: Chem.* 2001, *140*, 185-189.
[133] Klán, P.; Literák, J.; Hájek, M. *J. Photochem. Photobiol. A: Chem.* 1999, *128*, 145-149.
[134] Literák, J.; Klán, P. *J. Photochem. Photobiol. A: Chem.* 2000, *137*, 29-35.
[135] Klán, P.; Literák, J.; Relich, S. *J. Photochem. Photobiol. A: Chem.* 2001, *143*, 49-57.
[136] Chemat, S.; Aouabed, A.; Bartels, P. V.; Esveld, D. C.; Chemat, F. *J. Microwave Power E. E.* 1999, *34*, 55-60.
[137] Horikoshi, S.; Saitou, A.; Hidaka, H.; Serpone, N. *Environ. Sci. Technol.* 2003, *37*, 5813-5822.
[138] Horikoshi, S.; Tokunaga, A.; Hidaka, H.; Serpone, N. *J. Photochem. Photobiol. A: Chem.* 2004, *162*, 33-40.
[139] Horikoshi, S.; Kajitani, M.; Serpone, N. *J. Photochem. Photobiol. A: Chem.* 2007, *188*, 1-4.
[140] Horikoshi, S.; Hojo, F.; Hidaka, H.; Serpone, N. *Environ. Sci. Technol.* 2004, *38*, 2198-2208.
[141] Horikoshi, S.; Hidaka, H.; Serpone, N. *J. Photochem. Photobiol. A: Chem.* 2002, *153*, 185-189.
[142] Liu, Y.; Yang, S.; Hong, J.; Sun, C. *J. Hazard. Mater.* 2007, *142*, 208-215.
[143] Ai, Z.; Yang, P.; Lu, X. *J. Hazard. Mater. B* 2005, *124*, 147-152.
[144] Zhang, X.; Wang, Y.; Li, G. *J. Mol. Catal. A: Chem.* 2005, *237*, 199-205.
[145] Hong, J.; Sun, C.; Yang, S.-G.; Liu, Y.-Z. *J. Hazard. Mater.* 2006, *133*, 162-166.
[146] Matusiewicz, H.; Stanisz, E. *Microchem. J.* 2007, *86*, 9-16.
[147] Kataoka, S.; Tompkins, D. T.; Zeltner, W. A.; Anderson, M. A. *J. Photochem. Photobiol. A: Chem.* 2002, *148*, 323-330.
[148] Církva, V.; Žabová, H.; Hájek, M. *J. Photochem. Photobiol. A: Chem.* 2008, *198*, 13-17.
[149] Církva, V.; Relich, S.; Hájek, M. *J. Chem. Technol. Biotechnol.* 2010, in press (DOI: 10.1002/jctb.2281).
[150] Žabová, H.; Církva, V.; Hájek, M. *J. Chem. Technol. Biotechnol.* 2009, *84*, 1125-1129.
[151] Horikoshi, S.; Hidaka, H.; Serpone, N. *Environ Sci Technol* 2002, *36*, 5229-5237.

[152] Ai, Z.; Yang, P.; Lu, X. *Fresenius Environ Bull* 2004, *13*, 550-554.
[153] Ai, Z.; Yang, P.; Lu, X. *Chemosphere* 2005, *60*, 824-827.
[154] Wu, G.; Yuan, S.; Ai, Z.; Xie, Q.; Li, X.; Lu, X. *Fresenius Environ Bull* 2005, *14*, 703-708.
[155] Zhang, X.; Li, G.; Wang, Y. *Dyes Pigments* 2007, *74*, 536-544.
[156] Obee, T. N.; Hay, S, O.; Sangiovanni, J. J.; Hertzberg, J. B. *PCT Int. Appl.* WO 03094982 (2003).
[157] Lifongo, L. L.; Bowden, D. J.; Brimblecombe, P. *Chemosphere* 2004, *55*, 467-476.
[158] Chemseddine, A.; Boehm, H. P. *J. Mol. Catal.* 1990, *60*, 295-311.
[159] Ollis, D. F.; Hsiao, C.-Y.; Budiman, L.; Lee, C.-L. *J. Catal.* 1984, *88*, 89-96.
[160] Vincze, L.; Kemp, T. J. *J. Photochem. Photobiol. A: Chem.* 1995, *87*, 257-260.
[161] Mylonas, A.; Hiskia, A.; Papaconstantinou, E. *J. Mol. Cat. A: Chem.* 1996, *114*, 191-200.
[162] Kopf, P.; Gilbert, E.; Eberle, S. H. *J. Photochem. Photobiol. A: Chem.* 2000, *136*, 163-168.
[163] Mas, D.; Pichat, P.; Guillard, C.; Luck, F. *Ozone Sci. Eng.* 2005, *27*, 311-316.
[164] Li, Y.; Xie, Y.; Peng, S.; Lu, G.; Li, S. *Chemosphere* 2006, *63*, 1312-1318.
[165] Kolthoff, I. M. *Z. Anal. Chem.* 1922, *61*, 332-343.
[166] Herrmann, J.-M. *Catal. Today* 1995, *24*, 157-164.
[167] Terzian, R.; Serpone, N. *J. Photochem. Photobiol. A: Chem.* 1995, *89*, 163-175.
[168] Bouras, P.; Stathatos, E.; Lianos, P. *Appl. Catal. B* 2007, *73*, 51-59.
[169] Cramer, S. D. *Ind. Eng. Chem. Process Des. Dev.* 1980, *19*, 300-305.
[170] Vincze, L.; Kemp, T. J. *J. Photochem. Photobio.l A: Chem.* 1995, *87*, 257-260.

Index

A

B

C

D

E

L

M

N

O

P

Q

R

S